AS/A2 GEOGRAPHY

CONTEMPORARY CASE STUDIES

Climate Change

David Redfern

Series editor: Sue Warn

Philip Allan Updates, an imprint of Hodder Education, an Hachette UK company, Market Place, Deddington, Oxfordshire OX15 0SE

Orders

Bookpoint Ltd, 130 Milton Park, Abingdon, Oxfordshire, OX14 4SB
tel: 01235 827720
fax: 01235 400454
e-mail: uk.orders@bookpoint.co.uk

Lines are open 9.00 a.m.–5.00 p.m., Monday to Saturday, with a 24-hour message answering service. You can also order through the Philip Allan Updates website: www.philipallan.co.uk

ISBN 978-0-340-99184-8

First printed 2010
Impression number 5 4 3 2 1
Year 2015 2014 2013 2012 2011 2010

Printed in Italy

Hachette UK's policy is to use papers that are natural, renewable and recyclable products and made from wood grown in sustainable forests. The logging and manufacturing processes are expected to conform to the environmental regulations of the country of origin.

P01643

Contents

Introduction

Climate change is one of the most discussed issues of our time and our response to it continues to dominate the media. The climate of the Earth is always changing. In the past, it has altered as a result of natural causes. Today, however, the term climate change is generally used when referring to changes in our climate that have been identified since the early part of the twentieth century. The changes we have seen over recent years, and those that are predicted over the next 100 years, are thought by many to be largely the result of human behaviour, rather than because of natural changes in the atmosphere.

When talking about climate change, the **greenhouse effect** is important because it relates to the gases that keep the Earth warm. Although the greenhouse effect is a naturally occurring phenomenon, it is believed that the effect could be intensified by human activity, particularly the emission of gases into the atmosphere. It is the extra greenhouse gases that humans have released that are thought to pose the strongest threat, now known colloquially as '**global warming**'.

Scientists around the globe are looking at the evidence regarding climate change and using supercomputer models to come up with predictions for our future environment and weather.

The next stage of that work, which is just as important, is to look at the knock-on effects of potential changes. For example, are we likely to see an increase in precipitation and sea levels? Does this mean there will be an increase in flooding and, if so, what can we do to protect ourselves? Will other areas suffer from increased drought, and what can be done to rectify this? How will our health be affected by climate change, how will agricultural practices change and how will wildlife cope? While the whole concept may be controversial, some would argue that climate change could bring with it positives as well as negatives.

The list of things we need to think about that could be affected by climate change is endless. There are numerous examples of how we might change the way we live in order to cope with potential climate changes. The use of renewable energy is becoming increasingly popular. There are a number of possibilities for alternative energy sources including, for example, solar power and wind power.

In 1997, the Kyoto Treaty was set up to consider what could be done to reduce global warming. The treaty was established by the United Nations Framework Convention on Climate Change (UNFCCC) and involved most countries, but not the USA. The attempt to reach a new international settlement on greenhouse gas emissions to follow on from the Kyoto Protocol is seen as the single most important issue within the field of international climate change politics. The findings of the 2007 Intergovernmental Panel on Climate Change report reinforced the widespread consensus that action was required quickly to stabilise atmospheric greenhouse gas concentrations. Scientists are currently striving to present the most up-to-date research on all aspects of climate science to inform international negotiations that are expected to take place in Copenhagen at the end of 2009. The issue of climate change is here to stay.

About this book

Part 1 examines the reasons why the issue of climate change has become so controversial, both in the media and in the scientific community.

Part 2 deals with how climate change can be assessed across a variety of timescales from the longer term (geological), to the medium term (historical) to the short term (recent). It examines the evidence that suggests that climate change has occurred on each of these timescales.

Part 3 looks at the causes of climate change from astronomical forcing, to sunspot activity to the enhanced greenhouse effect of recent decades. Within this section, the impact of the short-term climate changes known as El Niño and La Niña are examined.

Part 4 is a large section that examines in detail, through case studies, a wide range of environments and locations that already show signs of stress caused by climate change, and where further impacts are likely. A full range of case studies is provided from the poles to the equator, from the permafrost to the rainforests, from the UK to small islands in the Pacific Ocean.

Part 5 introduces some of the work and views of the various bodies that have responsibility for looking at the causes and impacts of, as well as the possible solutions to, climate change. At a global scale, the most influential body is the Intergovernmental Panel on Climate Change (IPCC); in the UK, the *Stern Review* has had a significant input into government policy. The World Health Organization also needs to examine possible implications for its work.

Parts 6 and **7** are large sections that examine the various strategies suggested to address climate change. There are two main groups of such strategies: mitigation and adaptation. Through case studies, these sections look at a range of proposals in a range of countries and environments. One of the key areas of consideration is the difference between the abilities of human systems and natural systems to respond to both mitigation and adaptation. A considerable amount of research is being invested in both mitigation and adaptation, some of which is quite futuristic and, by definition, untried. However, many are aware of the business opportunities that they may provide. It is also important to recognise that mitigation strategies and adaptation strategies may be interrelated, and not always discrete.

Part 8 examines the range of views on climate change that exists. Even at the end of the first decade of the twenty-first century there is still no consensus on the existence of climate change, its possible impact, and the possible solutions to it.

Within this section, a selection of views from other organisations and individuals is presented and you are invited to either consider them, or to find out more about the protagonists and their views.

Part 9 concludes the book by considering some small-scale ways in which you could research and investigate climate change and attitudes to it.

Advice is given throughout on making the best use of case studies, both at AS and A2. (After some of the case studies, there are **Using case studies** boxes.) Most of these show how a particular case study might be useful for answering a specific question; some invite you to try your hand at an exercise based on one or more of the case studies. You will also find advice on tackling tasks that are commonly required in examinations.

Key terms

Adaptation: changing our lifestyles to cope with a new warmer environment, rather than trying to stop climate change.

Albedo: the amount of incoming solar radiation (insolation) that is reflected by the Earth's surface and atmosphere.

Carbon offsetting: a financial instrument representing a reduction in greenhouse gas emissions. For example, individuals, companies, or governments can purchase carbon offsets to mitigate their own greenhouse gas emissions from transportation, electricity use and other sources.

Carbon trading: a scheme that allows countries that have carbon emission units to spare — emissions permitted but not used — to sell this excess capacity to countries that are over their targets (as set by the Kyoto Protocol). One of the best known is EUETS (see below).

Clean coal: the colloquial name for Carbon Capture and Storage (CCS) which aims to take carbon out of the production of electricity at power stations by removing it with either a pre-combustion system or a post-combustion system.

DEFRA: the UK's Department for Environment, Food and Rural Affairs.

Dendrochronology: the analysis of tree rings from core samples, which can be used to provide evidence of past climates.

ENSO: the name given to the continually oscillating climate pattern in the Pacific Ocean that contains large circulations of warm and cold water. The phenomena known as El Niño and La Niña are linked to changes in these circulations.

EUETS: the European Union greenhouse gas Emission Trading Scheme. It came into effect in 2005 and is the largest multi-government trading scheme in the world.

Feedback mechanism: when the output of a system acts to amplify (positive feedback) or reduce (negative feedback) further output.

Global warming: the colloquial name given to the gradual warming of the Earth's atmosphere from the latter part of the twentieth century onwards. It is thought to be associated with the activity of humans. It is sometimes referred to as the enhanced greenhouse effect.

Greenhouse effect: the name given to the natural phenomenon whereby the Earth's atmosphere is warmed due to the trapping of heat by gases in the atmosphere such as carbon dioxide and water vapour, which would otherwise be radiated back into space.

Greenhouse gases: gases in the atmosphere, both natural and anthropogenic (caused by humans), which absorb outgoing long-wave radiation. They include carbon dioxide, chlorofluorocarbons (CFCs), methane, water vapour, nitrous oxide and ozone.

Ice core: a thread of ice that is obtained by drilling down through an ice sheet in order to examine the rings of annually accumulated snow. Ice cores can provide evidence of past climate change.

IPCC: the Intergovernmental Panel on Climate Change. It was formed jointly in 1988 by two United Nations bodies, the World Meteorological Organization and the United Nations Environment Programme. The organisation has produced four assessment reports on climate change.

Intermediate technology: the matching of technology to the needs and skills of the people of the area where it is to be used. It is a feature of many small-scale energy schemes in the developing world.

Kyoto Protocol: the outcome of a meeting in 1997 in which over 100 governments signed a 'Climate Change Protocol'. This set specific legally binding targets for pollution mitigation and proposed schemes to enable governments to reach these targets. Most governments agreed that by 2010 they should have reduced their atmospheric pollution levels of greenhouse gas emissions to those present in around 1990.

LULUCF (land use, land-use change and forestry): defined by the UN Climate Change Secretariat as 'a greenhouse gas inventory sector that covers emissions and removals of greenhouse gases resulting from direct human-induced land use, land-use change and forestry activities'.

Milankovitch cycles: changes in the surface temperature of the Earth due to variations in the orbit and axis tilt over time, which lead to changes in the amount and distribution of solar radiation received by the Earth.

Mitigation: the reduction in output of greenhouse gases and/or increasing the size and amount of greenhouse gas storage or sink sites.

Permafrost: ground in which the temperature has been below freezing for more than 2 years.

Renewable obligation certificate (ROC): a regime within the UK whereby energy suppliers face an increasing requirement year on year to supply a rising proportion of renewable energy, reaching 20% by 2020.

***Stern Review*:** a report commissioned by the UK government in 2006 to discuss the effect of climate change and global warming on the world economy, following the first three IPCC reports.

Sunspots: dark spots that appear on the Sun's surface, caused by intense magnetic storms. The effect of sunspots is to blast more solar radiation (insolation) towards the Earth.

Tipping point: the theoretical point after which the effects of climate change become irreversible.

UNFCCC: the United Nations Framework Convention on Climate Change agreed at the Earth Summit in Rio de Janeiro in 1992.

WHO: the World Health Organization.

Websites

www.ukcip.org.uk — The UK Climate Impacts Programme (UKCIP) helps organisations to adapt to climate change. UKCIP is mainly funded by the Department for Environment, Food and Rural Affairs (**DEFRA**). Other contributors include the Environmental Change Institute (Oxford University) and the UK Government's Knowledge Transfer Partnership scheme.

www.arctic-council.org — The Ottawa Declaration of 1996 formally established the Arctic Council as a high-level intergovernmental forum to provide a means for promoting cooperation, coordination and interaction among the Arctic States, with the involvement of the Arctic Indigenous communities and other Arctic inhabitants on common Arctic issues, in particular issues of sustainable development and environmental protection in the Arctic. Member States of the Arctic Council are Canada, Denmark (including Greenland and the Faroe Islands), Finland, Iceland, Norway, Russian Federation, Sweden and the USA.

www.antarctica.ac.uk — The British Antarctic Survey (BAS) is one of the world's leading environmental research centres, and is responsible for the UK's national scientific activities in Antarctica.

www.naturescalendar.org.uk — Nature's Calendar, from the Woodland Trust, is the home for thousands of volunteers who record the signs of the seasons where they live. It could mean noting the first ladybird or swallow seen in your garden in spring, or the first blackberry in your local wood in autumn. You could participate in their research.

www.bbc.co.uk/springwatch — This is the BBC's own version of the above, based on the popular BBC2 series. It contains a separate area called 'Bloom', which examines the theme of climate change, especially in terms of actions that can be taken by individuals.

www.metoffice.gov.uk/climate/uk/averages — The World Meteorological Organization (WMO) requires the calculation of averages for consecutive periods of 30 years, with the latest covering the 1961–1990 period. However, many WMO members, including the UK, update their averages at the completion of each decade. These averages help to describe the climate and are used as a base with which current conditions can be compared. Use the links to see a selection of station, district and regional averages or UK and regional maps for a wide range of weather elements.

www.ceh.ac.uk/data/nrfa/index.html — The Centre for Ecology and Hydrology's website can be used to explore records of over 50 000 individual years of daily and monthly river flow data, deriving from over 1300 gauging stations. These hydrological data underpin the sustainable exploitation and management of water resources and river systems in the UK.

http://royalsociety.org — The Royal Society is an independent scientific academy of the UK and Commonwealth dedicated to promoting excellence in science. It has published several reports on climate change, summarising current thinking on the issue.

www.tyndall.ac.uk — The Tyndall Centre for climate change research brings together scientists, economists, engineers and social scientists, who together are working to develop sustainable responses to climate change through inter-disciplinary research and dialogue on both a national and international level — not just within the research community, but also with business leaders, policy advisors, the media and the public in general.

www.foe.co.uk — Friends of the Earth is a major environmental pressure group. Here you will find a mass of information on climate change, but note that much of this is written from a particular viewpoint.

www.ipcc.ch — The official website of the Intergovernmental Panel on Climate Change. You will find full versions, as well as separate elements, of their Assessment Reports, including information on the process towards their Fifth Assessment Report. Try not to be overwhelmed by the sheer mass of information contained here.

Part 1

Why is climate change a controversial issue?

The climate change controversy concerns the nature, causes and consequences of **global warming**. The disputed issues include the causes of increased global average air temperature, especially since the mid-twentieth century, whether this warming trend is unprecedented or within normal climatic variations, and whether the increase is wholly or partially a result of human activity. Additional disputes concern estimates of climate sensitivity, predictions of additional warming, and what the consequences of global warming will be. The debate is vigorous in the popular media and at a policy level, with individuals, businesses, national governments and supranational organisations all being involved.

There has been a debate among public commentators about how much weight and media coverage should be given to each side of the controversy. Andrew Neil of the BBC stated that: 'There's a great danger that on some issues we're becoming a one-party state in which we're meant to have only one kind of view. You don't have to be a climate-change denier to recognise that there's a great range of opinion on the subject.' In June 2007, an Ipsos Mori poll conducted in the UK found that 56% of respondents believed that scientists were still questioning climate change. The survey also suggested that terrorism and crime were of more concern than climate change. A recent Canadian survey reported that the public has a poor understanding of the science behind global warming. This is despite recent publicity through various means, including the films *An Inconvenient Truth* and *The 11th Hour*. An example of this poor understanding is confusion between global warming and ozone depletion.

There are variations in opinions around the world. A 15-nation poll conducted in 2006 by Pew Global found that:

> There is a substantial gap in concern over global warming — roughly two-thirds of Japanese (66%) and Indians (65%) say they personally worry a great deal about global warming. Roughly half of the populations of Spain (51%) and France (46%) also express great concern over global warming, based on those who have heard about the issue. But there is no evidence of alarm over global warming in either the United States or China — the two largest producers of greenhouse gases. Just 19% of Americans and 20% of the

> Chinese who have heard of the issue say they worry a lot about global warming — the lowest percentages in the 15 countries surveyed. Moreover, nearly half of Americans (47%) and somewhat fewer Chinese (37%) express little or no concern about the problem.

On the other hand, a 47-nation poll conducted by the same organisation in 2007 found that: 'Substantial majorities in 25 of 47 countries say global warming is a "very serious" problem.'

The controversy concerning the science

Environmental groups, many governmental and non-governmental organisation (NGO) reports, and the non-US media often state that there is virtually unanimous agreement in the scientific community in support of human-caused global warming, although there is less agreement on the specific consequences of this warming. On the other hand, opponents maintain that most scientists either consider global warming unproven or dismiss it altogether, or they highlight the dangers of focusing on only one viewpoint in the context of unsettled science. Others maintain that either proponents or opponents have been stifled or driven underground.

The majority of climate scientists agree that global warming is caused primarily by human activities such as deforestation and burning fossil fuels. The conclusion that global warming is caused mainly by human activity, and will continue if greenhouse gas emissions are not reduced, has been endorsed by more than 50 scientific societies and academies of science, including all the national academies of science of the major industrialised nations and the Intergovernmental Panel on Climate Change (**IPCC**), established by the United Nations and World Meteorological Organization.

However, others have challenged the claim that scientific consensus has been reached. For example, Richard Lindzen wrote in the *Wall Street Journal* in 2006:

> Scientists who dissent from the alarmism have seen their grant funds disappear, their work derided, and themselves libelled as industry stooges, scientific hacks or worse. Consequently, lies about climate change gain credence even when they fly in the face of the science that supposedly is their basis.

Furthermore, some sceptics have compared the theory with a political dogma. Emeritus Professor Philip Stott wrote: 'Global warming has become the grand political narrative of the age, replacing Marxism as a dominant force for controlling liberty and human choices.'

There has been further controversy as each side of the debate has been accused of falsifying lists of supporters of its views, with several scientists being listed without either their knowledge or consent. There has been accusation and counter-accusation, much of which is linked to the availability of funding for further research, as well as funding to try to highlight strengths and weaknesses in the opposing view's scientific processes. Similarly, in some parts of the world, for example the USA, there have been accusations that scientific reports regarding climate change have been suppressed in order to play down its importance. Politics and science seem to be overlapping significantly in this context.

Political, economic and social aspects of the controversy

As more evidence has become available over the existence of global warming, debate has moved to further controversial issues, including:

- the possible economic, social and environmental impacts of climate change
- the appropriate response to climate change
- whether decisions on dealing with climate change require less uncertainty

The other point that leads to major controversy — because it could have significant economic impacts — is whether action (usually restrictions on the use of fossil fuels to reduce emissions of carbon dioxide) should be taken now or in the near future, and whether those restrictions would have any meaningful effect on global temperature. A strictly economic argument for or against action on climate change is, at best, limited because it fails to take into consideration other potential impacts (e.g. social and environmental) of any change.

The **Kyoto Protocol** is the most prominent international agreement on climate change and is also, in itself, controversial. Some argue that it goes too far, or not far enough, in restricting emissions of **greenhouse gases**. Another area of controversy is the fact that China and India, the world's two most populous countries, both ratified the protocol but, under the present agreement, are not required to reduce or even limit the growth of their carbon emissions. Nevertheless, China is the world's second largest producer of greenhouse gas emissions and India is the fourth. Various predictions see China overtaking the USA in total greenhouse emissions by 2010. The only major developed nation that has signed but not ratified the Kyoto Protocol is the USA. The countries with no official position on Kyoto are mainly African countries with underdeveloped scientific infrastructure, or are oil producers.

Another debate that is emerging is whether the world should adapt to climate change, or attempt to mitigate (moderate) it by, for example, reducing the consumption of fossil fuels. Once again economic factors come to the fore. Danish academic Bjørn Lomborg says:

> Despite our intuition that we need to do something drastic about global warming, we are in danger of implementing a cure that is more costly than the original affliction: economic analyses clearly show that it will be far more expensive to cut carbon dioxide emissions radically than to pay the costs of adaptation to the increased temperature.

Others argue that if developing nations reach the wealth levels of the USA, this could greatly increase consumption of fossil fuels and carbon dioxide emissions. In the next few decades as their economies grow, large developing nations such as India, China and Brazil are predicted to become major emitters of greenhouse gases.

In the early twenty-first century in the USA, the Bush administration supported an **adaptation**-only policy. While eventually recognising that most of the blame for recent global warming was due to human actions, it did not propose any major shift in the policy on greenhouse gases. Instead, it recommended adapting to inevitable changes, rather than making rapid and drastic reductions in greenhouse gases to limit warming. Some have found this attitude disingenuous and indicative of an inherent bias against prevention (i.e. reducing emissions/consumption) and for prolonging profits to the oil industry at the expense of the environment. UK journalist George Monbiot says:

> Now that the dismissal of climate change is no longer fashionable, the professional deniers are trying another means of stopping us from taking action. It would be cheaper, they say, to wait for the impacts of climate change and then adapt to them.

For many, the inauguration of Barack Obama in 2009 brought a new sense of optimism in the world — he is seen as a 'green' president.

Others have written of how the increasing use of words such as 'catastrophic' and 'irreversible' has affected the public discourse around climate change. One scientist, Mike Hulme, wrote:

> I have found myself increasingly chastised by climate change campaigners when my public statements and lectures on climate change have not satisfied their thirst for environmental drama and exaggerated rhetoric.

This book seeks to examine the main elements of the issue in a balanced manner: the extent of climate change, the reasons for this change, the consequences of climate change, and the ideas and practices to address it. The focus of this series is one of using up-to-date case studies to reinforce understanding. Hence, place-specific case studies are important. However, bearing in mind the range of opinion that exists, an important feature of this topic is that such case studies should also include varying viewpoints on this controversial issue.

Using case studies 1

Question

'Climate change is one of the most discussed issues of our time.' Explain why this is so.

Guidance

One of the most demanding skills you face on an A-level course, and one that continues into higher education, is the ability to present a balanced argument. Climate change, or global warming as it is often referred to in the media, is an issue that is causing a great deal of debate with strongly held views of protagonists on either side, and, in the middle. There is so much opportunity for debate:

- the science of its cause
- the possible impacts
- the possible solutions
- the decisions that may have to be taken at a variety of levels
- the language that is used by all sides in presenting their case
- whether 'Joe Public' actually cares

Before you embark on the rest of this book, compile a table (see example below) that summarises the views from both 'camps' — those who think it is of serious concern to the planet and those who think it is just 'hot air'. We will revisit this type of exercise at the end of the book.

Source of debate	Serious concern	Of little concern
The science		
The impacts		
The solutions		
The decisions etc.		

Part 2

How and when has climate changed?

Climate change can be assessed across a variety of timescales:

- long-term (or geological) — over several hundreds of thousands to millions of years
- medium-term (or historical) — within the last few thousand years
- short-term (or recent) — within the last few decades

Case study 1 LONG-TERM CLIMATE CHANGE

The best evidence for this comes from Greenland and Antarctic **ice cores**. Cores removed from ice sheets reveal layers going down through the ice. Each layer records a season of snowfall, buried and compressed by later falls.

The 3200 m East Antarctic core records the climate of the last 800 000 years. Air bubbles trapped in the ice contain atmospheric carbon dioxide and the ice itself preserves a record of oxygen isotopes. Figure 2.1 shows how low concentrations (180 parts per million (ppm)) of carbon dioxide occur naturally during glacial (cold) periods and high concentrations (280 ppm) during interglacial (warm) periods. It is clear that atmospheric carbon dioxide levels are higher now (380 ppm) than at any time for over half a million years.

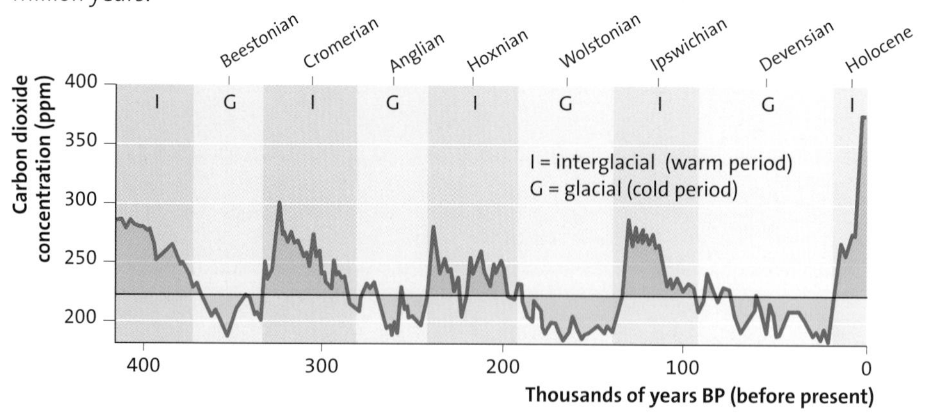

***Figure 2.1** Atmospheric carbon dioxide concentration measured from the Vostok ice core, also from East Antarctica*

Another method is to examine oxygen isotope levels in ocean-floor deposits. Core samples from the ocean floor reveal shifts in animal and plant populations, which indicate climatic change. The ratio of the isotopes oxygen-18 to oxygen-16 in calcareous ooze can also be measured. During colder phases, water evaporated from the oceans and precipitated onto the land eventually forms glacial ice. Water containing the lighter oxygen-16 isotope is more easily evaporated than that containing heavier oxygen-18. As a result, the oceans have a higher concentration of oxygen-18, while the ice sheets and glaciers contain more of the lighter oxygen-16. During warmer periods, the oxygen-16 held in the ice is released and returns to the oceans, balancing out the ratio. Studies of isotope curves showing the ratio of oxygen-16 to oxygen-18 therefore give a picture of climate change.

Recent investigations have suggested that isotope variations are an indication of changes in the volume of ice, rather than water temperature. However, as ice volume itself reflects climatic conditions, such studies have tended to confirm earlier findings.

Scientists have found that the results of climate fluctuations identified by both these methods correlate well. Therefore, climate change has been a feature for a very long period of time.

MEDIUM-TERM CLIMATE CHANGE

Case study 2

Studies related to changes in vegetation provide strong evidence of climate change over this time period.

Pollen extracted from sediment cores in peat bogs and lake beds records the ecology of the past. Pollen grains are preserved in waterlogged sediments, which are anaerobic (oxygen-free). Each plant species has a distinctively shaped pollen grain that can be identified. In the UK, pollen sequences have shown that ecosystems have changed in response to climate change. Tundra ecosystems were present in past glacial periods, whereas forest gradually colonised areas as interglacial conditions developed (see Table 2.1). However, long pollen sequences are rare and vegetation change may 'lag' behind climate change.

Dendrochronology is the analysis of tree rings from core samples. Each year, the growth of a tree is shown in its trunk by a single ring made up of two bands:

- a band reflecting rapid spring growth when the cells are larger
- a narrower band of growth during the cooler autumn

Table 2.1 *Medium-term climate change*

Climatic period	Time before present (years)	Climatic conditions
Sub-Atlantic to present day	0	A period of warming in the last 200 years A marked cool period between AD 1500 and AD 1800 — the Little Ice Age (Figure 2.2)
	2500	A cool coastal climate with cooler summers and increased rainfall
Sub-Boreal	5000	Period known as the neoglacial in Europe, with evidence of ice advance in alpine areas Warm summers and cold winters
Atlantic	7500	Temperatures reach the optimum for many trees and shrubs — a warm and wet climate
Boreal	9000	A continental-type climate — warmer and drier in summer, cold in winter
Pre-Boreal	10 300	Changing from tundra/subarctic conditions to more continental — mainly cold and wet

Figure 2.2
A frost fair on the Thames in 1677

The width of the ring depends on the conditions of that particular year. A wide band indicates a warm and wet year; a narrower one indicates cooler and drier conditions. The change in width from one year to another is of greater significance than the actual width, as bigger growth rings tend to be produced in the early life of the tree, irrespective of the conditions.

Recent investigations have shown that trees respond more to levels of moisture than to temperature. Dendrochronology has a limitation, in that few trees exist that are older than about 4000 years. It has been possible to extend surveys further back using remains of vegetation preserved in anaerobic conditions.

Historical proxy records can also be used to reconstruct climate before the start of instrumental records. These include paintings, diaries and journals that record weather. The Little Ice Age from AD 1500 to AD 1800 is recorded by a number of such sources.

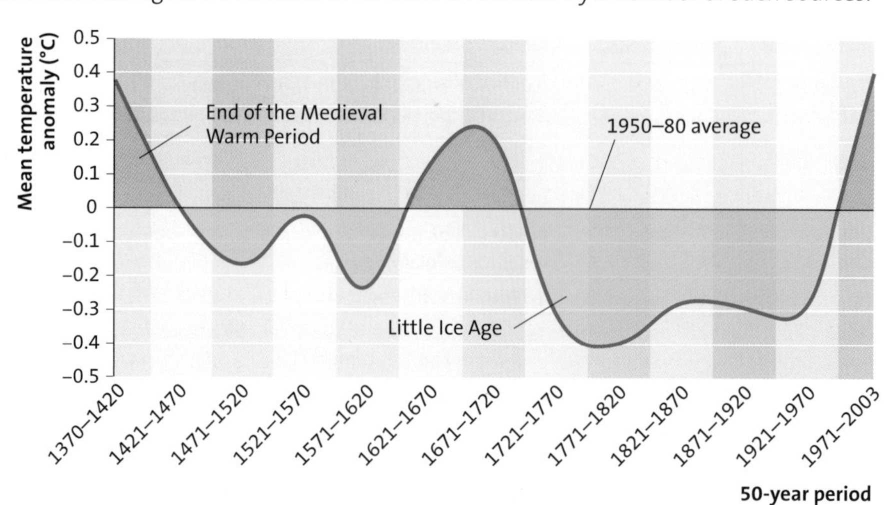

Figure 2.3
A climate proxy: temperature anomalies reconstructed from grape harvest dates, Burgundy, France, 1370–2003

Since the mid-fourteenth century, the date of the grape harvest in the Burgundy region of France has been recorded. This sequence of grape ripening dates is a proxy record and has been used to indicate past climate (see Figure 2.3). The graph shows evidence of significant warming since 1900 as well as the Little Ice Age. Proxy records, however, need to be used with care. The date of the grape harvest could have been affected by non-climate factors such as conflict in the area or disease. It is also difficult to know which of temperature, rainfall or sunshine hours may have had a greater influence on harvest dates.

Using case studies 2

Question

Ice cores, oxygen isotopes in ocean-floor deposits, pollen grains in peat bogs, dendrochronology and historical records (including paintings and literature) all provide evidence of climate change over the medium to longer term.

Choose three of these, describe what they say about climate change, and assess their reliability.

Guidance

There are two elements to this question. First, you should describe what the chosen indicators say about climate change — in other words, how the variations in the indicators are reflected in the interpretation of what climate was like in the past. This may be relatively straightforward — for example, in dendrochronology the warmer and wetter the climate, the larger the tree ring. However, to understand some of the other indicators you have to think carefully about what the evidence points towards. It is not enough in a question like this to say how the scientific test, or indicator, operates — you have to link the outcome to climate change.

The second element of the question asks you to consider the reliability of the outcome of a scientific test, or indicator. All scientific tests have a statement of error built into them. For literature and art, you have to consider other aspects that may have had a bearing. For example, was the author trying to embellish something? Or, was an artist carried away with his/her sense of creativity? Remember, such people did not set out to record climate for people in the twenty-first century and, also, the events portrayed may have been local to that area.

SHORT-TERM (RECENT) CLIMATE CHANGE

Case study 3

In 2007, the Intergovernmental Panel on Climate Change (IPCC) published its fourth Assessment Report (AR4). It stated: 'Warming of the climate system is unequivocal, as is now evident from observations of increasing global average air and ocean temperatures, widespread melting of snow and ice and rising global mean sea level.'

Instrumental records from weather stations have existed for the last 100 years or so. They show that near-surface air temperatures rose by 0.74°C between 1906 and 2005. The warming trend has been almost constant since 1960, and 11 of the world's hottest years since 1850 occurred between 1995 and 2006 (Figure 2.4).

Furthermore, the oceans have warmed to depths of 3000 m and sea levels have risen. Records show that global sea level has risen at the average rate of 1.8 mm per year (mm y^{-1}) between 1961 and 2003, with a faster rate (3.1 mm y^{-1}) in recent years.

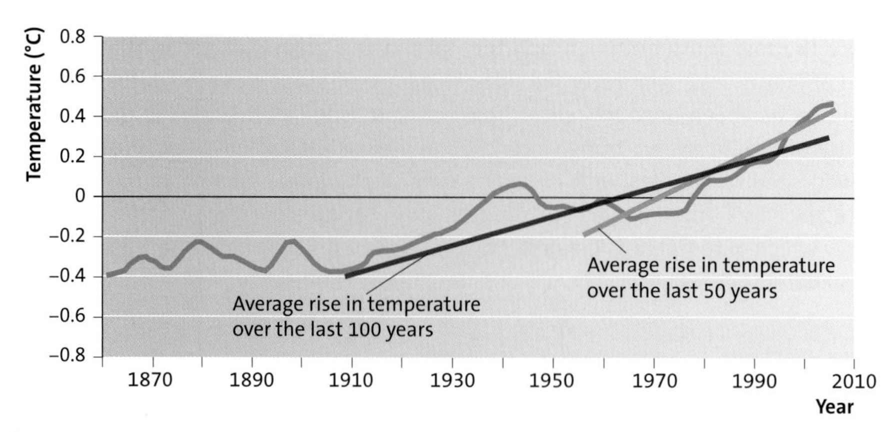

Figure 2.4
Global mean temperature change
Source: IPCC 2007

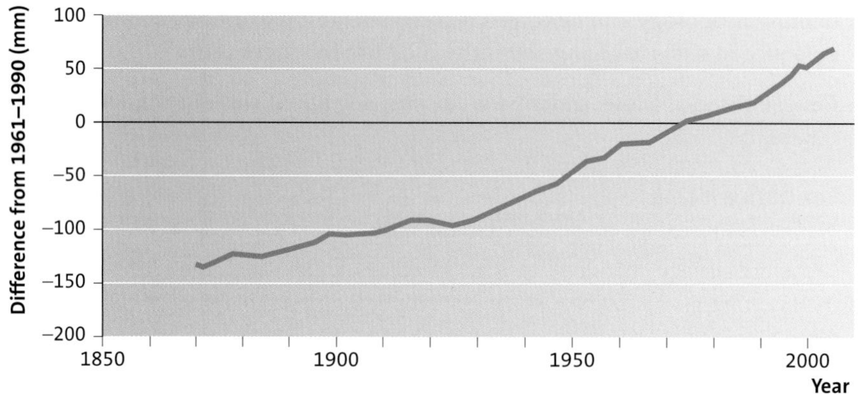

Figure 2.5
Global average sea level
Source: IPCC 2007

Most of this rise is attributed to thermal expansion, with water from melting glaciers and ice caps having a lesser impact (Figure 2.5).

Ice and snow levels are also key indicators of climate change (Figure 2.6). Some valley glaciers around the world have shrunk in size (Figure 2.7(a)), and there have been other significant changes.

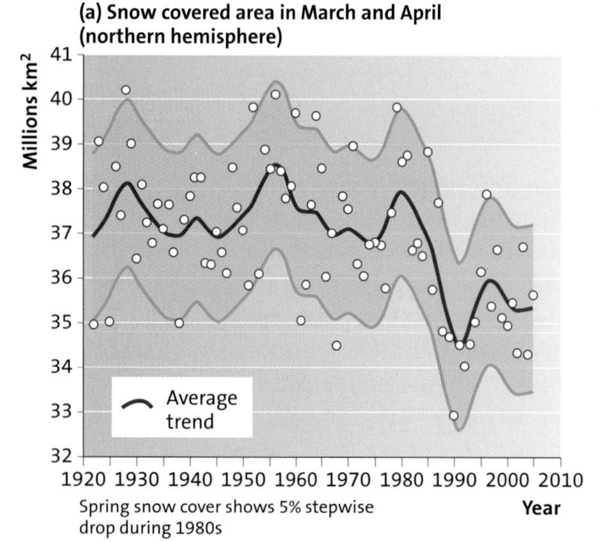

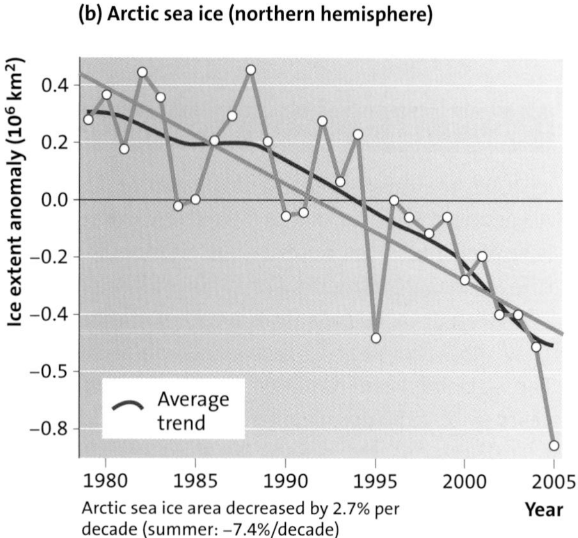

Figure 2.6
Ice and snow levels
Source: IPCC 2007

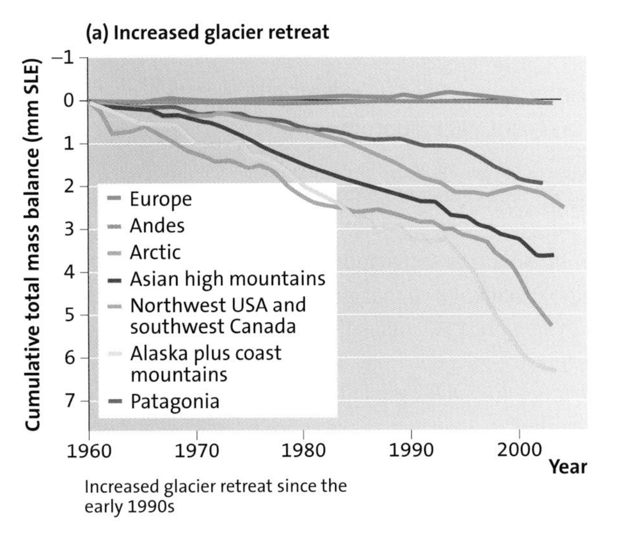

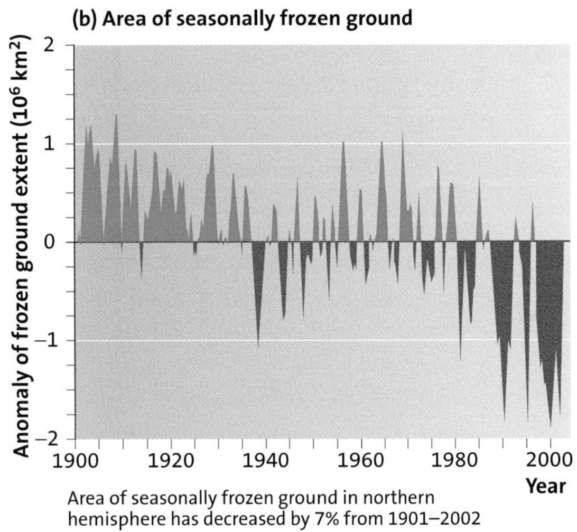

Figure 2.7 *Glacier retreat and seasonally frozen ground*
Source: IPCC 2007

- Annual average Arctic sea ice extent has shrunk by 2.7% per decade, with decreases in summer of 7.4%.
- Temperatures at the top of the **permafrost** layer have increased since the 1980s by up to 3°C.
- Since 1900, the maximum area covered by seasonally frozen ground has decreased by 7% in the northern hemisphere (Figure 2.7(b)).

Melting of Greenland ice sheet has increased by 16% since 1979. NASA satellites measured record melting in 2006: 239 km^3 of ice. Many glaciers, which drain the ice sheet, have doubled speed of flow since late 1990s

NASA satellite data show floating Arctic sea ice declining by 8.5% per decade. Arctic Ocean could be ice free in summer by 2060 if trends identified since 1979 continue

Evidence of melting ice in Antarctica less clear than for Arctic. Some small ice shelves in Antarctic Peninsula have collapsed: 1600 km^2 Larsen A in 1995, 1100 km^2 Wilkins in 1998 and 13 500 km^3 Larsen B in 2002. In 2005 British Antarctic Survey found that 85% of Antarctic Peninsula glaciers had retreated by an average of 600 m since 1953

World Glacier Monitoring Service reported in 2007 that 30 valley glaciers in nine mountain ranges were melting three times faster than in the 1980s. Many valley glaciers are 50% smaller than in 1850. Since 2000, Alpine glaciers have thinned by an average of 1 m per year. Similar rates observed in Andes, Patagonia, the Cascades range and Himalayas

Valley glaciers thinning by more than 1 m yr^{-1}

Figure 2.8 *Recent changes in global ice cover*

Question

Study Figures 2.4 to 2.8, which illustrate the evidence provided by the IPCC in its AR4 to establish that climate change has taken place in recent years. Comment on the strength of the evidence.

Guidance

The figures provided offer a significant amount of evidence to suggest that climate change and, in particular, warming, is taking place at a global scale. Note that the command words in the question do not ask you to describe the evidence (in most cases the evidence is simple to describe); it asks you to comment on it. 'Comment on' is a command that is becoming increasingly common in examination papers at A-level. It is asking you to examine the data, and then to make a statement that arises from the data that is relevant, appropriate and geographical, but above all *not* directly evident. This command is asking you to 'think like a geographer' and thereby demonstrate a degree of synthesis.

To illustrate this, two examples are given here:

Figure 2.4 shows that the warming of the planet has been most significant within the last 12 years to 2006, and two lines have been inserted to show the average rise in temperatures over the last 50 and 100 years. A commentary would be that these two lines have been chosen deliberately to show a trend and to make it appear more alarming. If the trend line had started in the 1870s for instance, the increase in temperatures would not have appeared as steep. Is it even correct to make judgements over such a short time period, bearing in mind the age of the planet?

Figure 2.7 shows the decrease in mass balances of glaciers in the latter part of the twentieth century. The inference is that glaciers are shrinking, and that this is evidence of global warming. However, the evidence is not clear cut — glaciers in Europe and the Andes appear not to be shrinking significantly compared with those in other parts of the world. Another point is that the graph does not show glaciers in Australasia, for example in New Zealand, where they have *increased* in size in recent decades. Again, is the evidence selective?

Part 3

The causes of climate change

There is no single cause of climate change. On the very long timescales of glacial to interglacial cycles, the most common explanation is the variation in the Earth's orbit around the Sun. On medium timescales of hundreds to thousands of years, variations in the Sun's solar output may fit the observed trends. The warming that the Earth has experienced in the last few decades (known as global warming) is seen increasingly as being driven by atmospheric pollution by humans (known as anthropogenic warming). In the IPCC's AR4 (2007) it is stated that: 'The understanding of anthropogenic warming and cooling influences on climate has improved since the Third Assessment Report (2001) leading to very high confidence that the globally averaged net effect of human activities since 1750 has been one of warming.' It also stated that 'most of the warming is very likely (odds 9 out of 10) due to greenhouse gases'.

Astronomical forcing

Milutin Milankovitch developed the theory of astronomical climate forcing in 1924. He argued that the surface temperature of the Earth changes over time because the orbit and axis tilt of the Earth vary over time. These variations lead to changes in the amount and distribution of solar radiation received by the Earth from the Sun. Over a timescale of 100 000 years, the Earth's orbit changes from circular to elliptical and back again. This affects the amount of radiation received from the Sun. On a timescale of 41 000 years, the Earth's axis tilts from 21.5° to 24.5° and back again. This changes the seasonality of the Earth's climate. The smaller the tilt, the smaller is the difference between summer and winter. In addition, on a 22 000-year timescale, the Earth's axis 'wobbles' and this changes the point in the year when the Earth is closest to the Sun.

In support of Milankovitch's theory is the fact that ice ages (glacials) have occurred at regular 100 000-year intervals. However, the actual impact of orbital changes on amount and distribution of solar radiation is small — probably only enough to change global temperature by 0.5°C. It is known from the evidence of past climate change that ice ages were about 5°C colder than interglacials.

Many scientists argue that **Milankovitch cycles** may have been enough to trigger a major global climate change, but that other mechanisms are needed to sustain it.

These mechanisms are known as 'feedback' effects. For example, small increases in snow and ice on a surface raise the **albedo** (the proportion of heat reflected) of that surface. More solar energy is reflected back by the white surface, contributing to further cooling, which in turn may encourage more snowfall. A small change therefore has an ever increasing effect.

Changes in solar output

The amount of energy emitted by the Sun varies as a result of **sunspots**. These are dark spots that appear on the Sun's surface, caused by intense magnetic storms. The effect of sunspots is to blast more solar radiation towards the Earth. There is a well-known 11-year sunspot cycle, as well as longer cycles. The total variation in solar radiation caused by sunspots is about 0.1%. Sunspots have been recorded for around 2000 years and there is a detailed record for around 400 years.

A long period with almost no sunspots occurred between 1645 and 1715, and this is linked to the Little Ice Age. Prior to that there was more intense sunspot activity, which could have led to the warmer conditions that existed at that time. Some scientists have suggested that around 20% of twentieth-century warming could be attributed to solar output variation.

TopFoto

Figure 3.1
Sunspots

The greenhouse effect

The **greenhouse effect** is a natural phenomenon. Atmospheric gases in the troposphere allow incoming short-wave radiation from the Sun to pass through and warm the Earth. Some of this radiation is reflected back at a longer wavelength from the Earth's surface into space. Greenhouse gases in the troposphere, such as water vapour and carbon dioxide, absorb some of this long-wave radiation and radiate it back again to the Earth's surface. This trapping of heat is known as 'the greenhouse effect' and is part of the natural process of heat balance in the atmosphere. It is essential for life on Earth. Temperatures are raised to a global average of 15°C; without the greenhouse effect the planet would be about 30°C colder. The greenhouse gases responsible for trapping heat include carbon dioxide, chlorofluorocarbons (CFCs), methane, water vapour, nitrous oxide and ozone.

Provided the amount of carbon dioxide and water vapour in the atmosphere stay the same and the amount of solar radiation is unchanged, then the temperature of the Earth remains in balance. However, this natural balance has been influenced by human activity. The atmospheric concentration of carbon dioxide has increased by about 15% in the last 100 years and the current rate of increase is estimated to

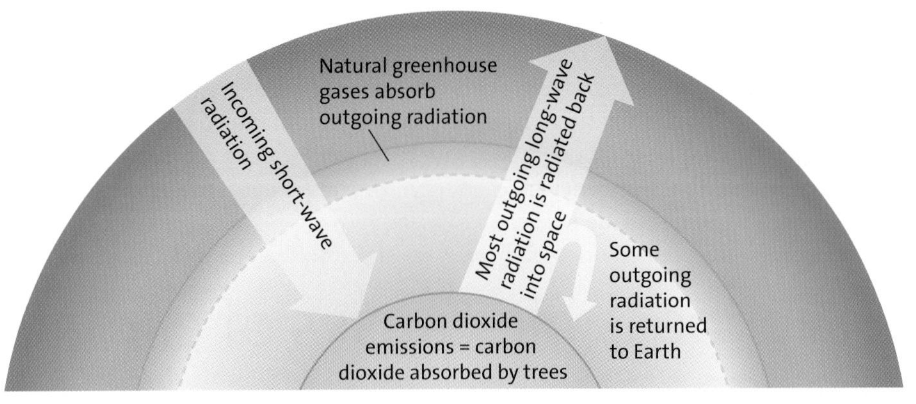

Figure 3.2
The greenhouse effect

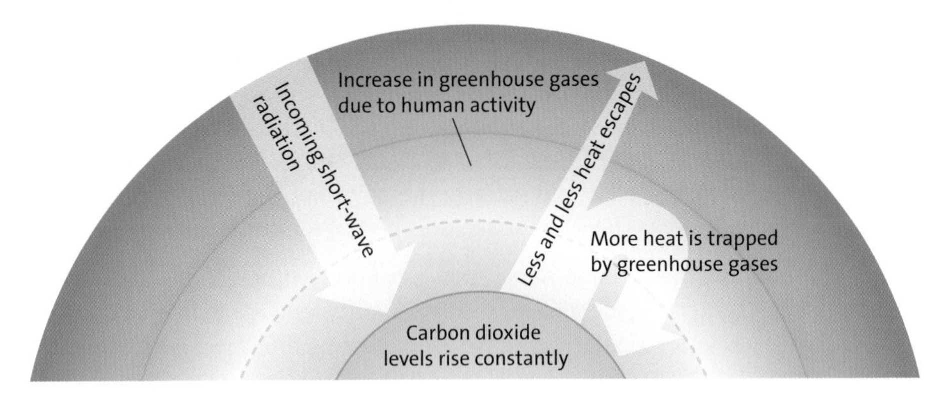

Figure 3.3
The enhanced greenhouse effect

be 0.4% per year. This, together with increases in levels of other greenhouse gases, such as methane and nitrous oxide, has upset the natural balance and led to global warming. Some scientists refer to these increased levels of greenhouse gases as the enhanced greenhouse effect.

In general, it is agreed that these continuing atmospheric changes will lead to a further rise in temperature. However, it is difficult to predict the extent or speed of change. If carbon dioxide levels double, then temperatures could rise by a further 2–3°C, with greater rises at higher latitudes, perhaps in the order of 7–8°C. It is believed that this warming will cause sea levels to rise.

One of the main reasons for the increase in atmospheric carbon dioxide has been the burning, in the industrialised nations of the world, of fossil fuels such as coal, oil and natural gas, which contain hydrocarbons. Developing countries (and some countries such as China are now well established) are beginning to generate energy as cheaply as possible. At the present time this means consuming huge quantities of fossil fuels, thereby adding to the problem.

Deforestation has also been linked to global warming. The rainforests of the world act as a 'carbon sink' because trees are a major store of non-atmospheric carbon. The more vegetation there is, the more carbon dioxide that can be processed by trees and other plants. The tropical rainforests are diminishing rapidly due to the demand for space and resources created by economic development in many countries — for example Brazil. Continued deforestation will, therefore, contribute to the build-up of carbon dioxide in the atmosphere. Large-scale pastoral farming, where huge herds of domestic cattle are reared for meat in areas cleared of forest, has been found to result in increased emissions of methane gas. Ironically, where the rainforests have been flooded to create reservoirs for the production of hydro-electric power (a form of renewable energy), decomposing vegetation within the lakes adds to carbon dioxide levels.

Other possible causes of climate change

Volcanic activity can alter global climate. Major eruptions eject material into the stratosphere where high winds distribute it around the world. Volcanoes eject huge volumes of ash, sulphur dioxide, water vapour and carbon dioxide. High in the atmosphere, sulphur dioxide forms a haze of sulphate aerosols that reduces the amount of sunlight received at the surface, thereby lowering temperatures. These changes are short lived as the sulphur aerosols persist for only 2 to 3 years.

Global dimming is a new phenomenon that has entered the debate. Atmospheric pollutants consisting of suspended particulate matter and sulphur dioxide reflect solar energy back into space and so have a net cooling effect. So, with the green-house effect having a warming effect, it is possible that human-created pollution is both warming and cooling the planet at the same time and that some pollutants are reducing the full impact of global warming. In both North America and Europe, soot and sulphur dioxide pollution has fallen dramatically since 1990 as a result of attempts to reduce acid rain. Could this have a link to the global warming trend that has accelerated since that time?

Case study 4 THE EL NIÑO SOUTHERN OSCILLATION (ENSO)

The Pacific Ocean contains large circulations of warm and cold water. El Niño and La Niña are linked to changes in these circulations, which in turn are linked to changes in atmospheric processes. This continually oscillating climatic pattern is known as the El Niño Southern Oscillation (**ENSO** — see Figure 3.4).

In a normal year, transfers of heat energy take place in the usual way between the equator and the subtropical high pressure zone and there are east-to-west surface circulations of warm air and water, called Walker circulations. The normal Walker circulation produces easterly trade winds between South America and Indonesia/ Australia, taking warm water in a westerly direction and allowing cold water to move north along the Pacific coast of Peru.

Every few years this Walker circulation breaks down. The easterly trade winds decline and warm water moves eastwards across the Pacific to shut off the cold Peru ocean current and produce a warm ocean current (El Niño, 'the Christ child') off the South American coast. The high pressure which normally forms over the cold ocean is replaced by low pressure over the warmer ocean, which can have a temperature

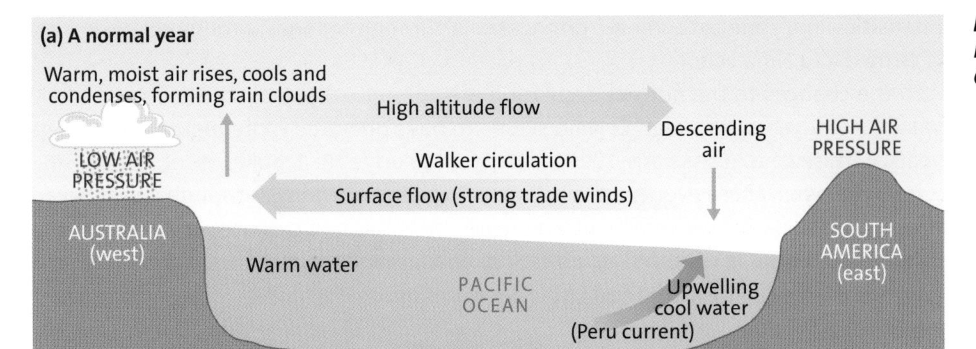

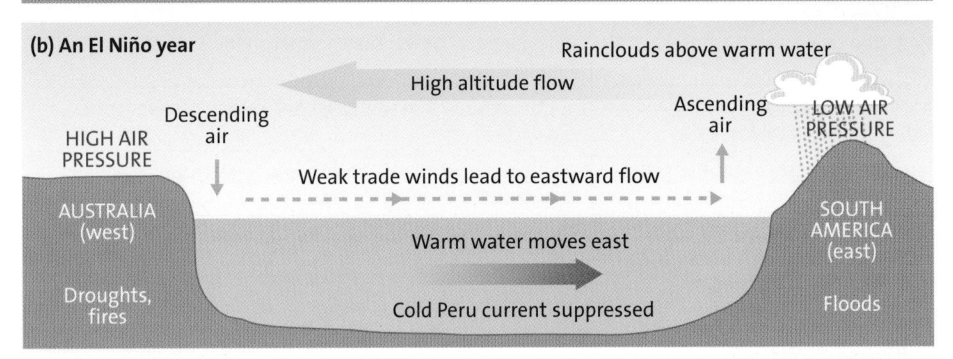

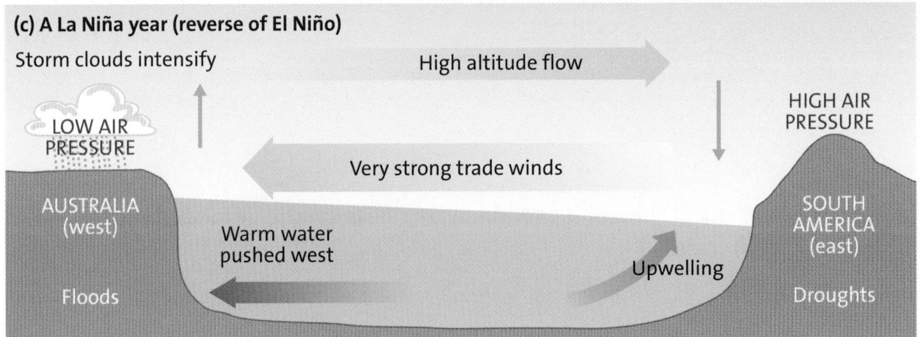

Figure 3.4 *El Niño Southern Oscillation*

6–10°C above normal. This produces heavy rainfall on the usually arid coastline of Peru and brings natural hazards, such as flooding. The change in circulation can have an effect elsewhere: the effect on the route of the jet stream can produce drought in other areas and even affect climatic events in the northern hemisphere, bringing increased snow and floods to North America. Conversely, the number of hurricane events in the Caribbean tends to be reduced in El Niño years.

Under the effects of the normal Walker circulation, the easterly trade winds cause warm water to build up in the area of the Coral Sea (between Australia and Indonesia). Sea levels rise and can cause ecological and economic damage to low-lying islands within the region. Recent research in Australia indicates that even a well-developed Walker circulation can cause extreme problems. With unusually strong easterlies, the cool water off South America can be moved northwards across the central Pacific, lowering ocean temperatures, pushing the warm water even further west and changing atmospheric pressure patterns to produce higher than average rainfall and cyclone intensities in a zone extending from Australia through Indonesia to Bangladesh. This is termed a La Niña event.

These changes to the normal circulation are not annual events; they occur at intervals of 2 to 7 years, although El Niño seems to have operated more regularly in recent years, with several in the early 1990s and a particularly catastrophic event in 1997–98. It is not known if these events are also linked to global warming, although it is believed that increased global warming could increase the variation attributed to ENSO. On the other hand, ENSO is regarded as a distinct phenomenon, and hence not all extreme weather events can be attributed to global warming.

Table 3.1
Effects of El Niño and La Niña events

During an El Niño event	During a La Niña event
Rainfall is reduced in southeast Asia, New Zealand, Australia and India leading to drought, crop failure and wildfires	Rainfall is higher than normal in Indonesia and the Philippines and lower than normal on the west coast of South America
Heavy rain in California, Mexico and the coasts of Peru and Ecuador often results in flooding and mudslides	Southern Africa and southeast Australia may experience floods
Suppression of the cold current in the east Pacific devastates fish catches off the west coast of South America	North eastern Africa, California and western South America may experience drought
Unusually strong winds in the Atlantic shear off the tops of clouds preventing convection cells forming, so there are fewer severe hurricanes in the USA and Caribbean	There are more hurricanes in the Caribbean and the southern USA
Tornadoes in the USA are reduced There are more cyclones in Hawaii and Polynesia, but fewer in northern Australia Southern Africa may experience drought while there may be floods in east Africa	

4 *Using case studies*

Question

Describe the varying effects of the El Niño and La Niña events both on the immediate areas affected and also on the global scale.

Guidance

The reasons for the ENSO (El Niño Southern Oscillation, including the event known as La Niña) are not yet understood fully, although their effects are described in many geographical texts. It would not be appropriate therefore to ask a question on their causes. However, one interesting aspect of this area of work is that both El Niño and La Niña affect not only their regional areas, but are also thought to affect weather and climate on a much bigger scale. This introduces another element of synopticity — the ability to see connections between different *scales* of effects: regional and global.

A response to this question should focus on the effects on two scales. For example, for La Nina:

Regional scale In northeastern Australia, there are warmer temperatures, lower atmospheric pressure, windier conditions, heavier rainfall, flash floods.

Global scale Rainfall is higher in Indonesia and lower on the west coast of South America; higher levels of rainfall on the eastern coast of southern Africa; lower levels of rainfall in northeastern Africa, California and western South America. Some people believe that hurricanes are more frequent in the Caribbean and southern USA.

Part 4

The impacts of climate change

The distribution of temperature and rainfall changes means that climate change will affect different geographical locations and environments in different ways. The following set of case studies examines the impact of climate change on a wide range of environments.

THE ARCTIC AND GLOBAL WARMING

Case study 5

In its fourth assessment report (AR4) published in 2007, the IPCC provided clear evidence regarding the relative speed of warming in the Arctic. The global mean temperature measured at the surface of the planet between 1906 and 2005 increased by 0.74°C. However, surface air temperatures north of 65°N have been warming at roughly twice the global rate for the last four decades (1965–2005).

The effects of this rapid warming in Arctic regions have been identified in numerous elements of the cryosphere (the planet's snow and ice), with knock-on effects on other parts of the environment. As expected, most elements of the cryosphere have been shrinking in recent times: snow cover has decreased in many regions, the extent of Arctic sea ice has decreased (in all seasons, but most dramatically in spring and summer), as has its thickness, and there was an almost worldwide retreat and melting of glaciers.

These trends are also affected by other factors. The changes in the Arctic ice reflect changes of polar winds, which determine subsequent changes in snowfall. For example, the change in the volume of glaciers has been calculated over the last four decades for the Arctic as a whole, and also for the three large subregions that make up the Arctic. Overall, from 1961 to 1998, the Arctic system of glaciers experienced a loss in ice volume of 400 km^3. Of the three subregions, large losses of ice were recorded for glaciers in the Canadian Arctic and the Russian Arctic, but glaciers in the European Arctic showed large increases in volume because glaciers in Scandinavia and Iceland gained, due largely to greater winter snowfall.

Effects on sea ice

Sea ice is the frozen crust of the ocean. It forms when air temperatures are low enough to freeze sea water (about –1.5°C) and where waves and wind do not break up the ice layer. Because of winds, tides and wave swell, sea ice may break into plates or floes, which can be up to several kilometres across. Floes are separated by areas of open water called leads (smaller) or polynyas (large). Where floes are driven into one another by winds or currents their edges can be buckled upwards to form pressure ridges, or downwards to form keels. Sea ice formed in a single winter is usually 2–4 m thick, but in many parts of the Arctic sea ice survives the summer by producing much thicker layers.

Sea ice is a critical element in the global weather system (Figure 4.1). Ice has a high albedo of 80%, and this is a major reason why the poles remain cool. Loss of sea ice can have a disproportionate effect because of this albedo. If a small amount of sea ice is lost, the newly exposed ocean absorbs more heat, warms up and is therefore likely to melt more sea ice, opening up more ocean and so on. This positive feedback process means that a small change in the extent of sea ice can lead quickly to enormous changes in climate and longer sea-ice change. In theory, this could lead to an ice-free Arctic.

The extent of sea ice is best studied from satellites that monitor changes over many thousands of square kilometres throughout the year, including times when ground-based measurements would not be possible. Until recently, the sea-ice extent in September was reducing by about 4% per decade. However, this trend has been accelerating and record minima in the last few years mean that the rate is now closer to 8% loss per decade. Overall, there has been a reduction in extent from about 12.6 million km^2 in 1979 to less than 11 million km^2 in 2006.

Scientists are not only concerned about the *extent* of the sea ice, but also about its *thickness*. Ever since the early days of the Cold War, nuclear submarines have been

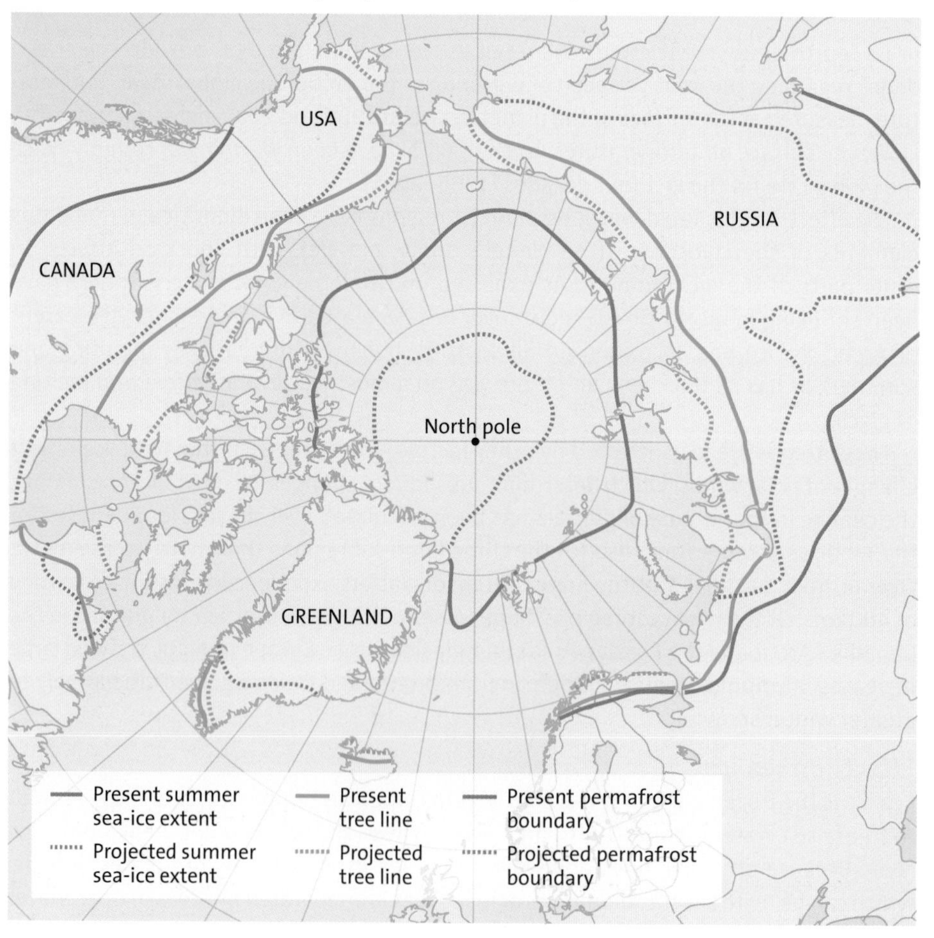

Figure 4.1 ***Arctic region: summary of key changes***

venturing under the ice. These submarines routinely measured the thickness of the ice above them. These measurements have been declassified recently, so researchers have been able to use them to study changes over recent decades. It is now clear that the Arctic ice thinned by an average of 1.3 m between the 1960s and the 1990s. This compares with an average thickness of 4 m at the beginning of this period.

Thinning is of significant concern because once the ice reaches a critical thinness it will break up and disintegrate, meaning that large areas of sea ice may suddenly become open ocean, reducing albedo further.

The combination of the retreat of the ice edge and the thinning of the remainder has led many scientists to predict an ice-free Arctic Ocean before the end of the century.

Effects on the natural world

Arctic ecology is affected by these changes in numerous ways. The warming of the atmosphere, which leads to warmer soils, seems to be having an effect on the Arctic tree line — the boundary between the boreal forest and the tundra. Evidence for an advance northwards of the tree line has been found in Alaska — studies of the age of trees along transects that cross the tree line from the forest to the tundra show that they are becoming progressively younger. This indicates tree-line advance, with a consequent movement north of associated species, e.g. lichens, mosses, fungi and birds such as woodpeckers.

Changes in the duration and timing of the growing season have also been detected, although patterns vary geographically. Studies using satellite images indicate that the length of the growing season is increasing by 3 days per decade in Alaska and 1 day per decade in northern Eurasia. However, there has been a delayed onset of the growing season in the Kola Peninsula in northern Russia, during a cooling of the climate over the past 2 years.

The loss of the sea ice may also affect the habitat of seals, polar bears, fish and other animals. Problems for polar bears have been recorded already. They have to wait longer each year for the sea ice to form, deal with diminished hunting grounds and undertake exhausting open-water swims to find ice, sometimes with tragic results. As the waters warm, numbers of some fish species such as the arctic char may decline; other fish, for example cod and herring, may increase. Land-based species that are adapted to the Arctic climate, such as lemming, arctic fox, snowy owl and caribou are also at risk.

Effects on human activities

The Arctic is not densely populated, but is home to 4 million people. The changes described above will affect their way of life. At sea, the decline in the Arctic Ocean's ice cover should facilitate marine transport and oil and gas exploration, but it is also likely to mean increasing dangers from icebergs.

On land, most of the Arctic is underlain by permafrost. The thawing of the permafrost has important implications for roads, buildings and pipelines that have been designed specifically for permanently frozen ground. Reinforcement of these structures and new engineering design and construction techniques are likely to be necessary in many areas.

Some of the Arctic's indigenous communities, for example the Inuit (Figure 4.2), continue to practise subsistence hunting; climate change is likely to have an impact on their harvest of animals. The numbers and distribution of walrus and seals are influenced strongly by the weather, particularly in the spring, and by the patterns of sea ice.

Figure 4.2 Inuits will be badly affected by Arctic climate change

Corel

People who depend on the herding of caribou will also be affected as the animals will have to cope with changes that include thermal stress due to warming and new migration patterns in response to new distributions of food and weather.

There will also be positive impacts — trade and tourism opportunities will increase (although some may regard this as a negative impact), and therefore incomes and

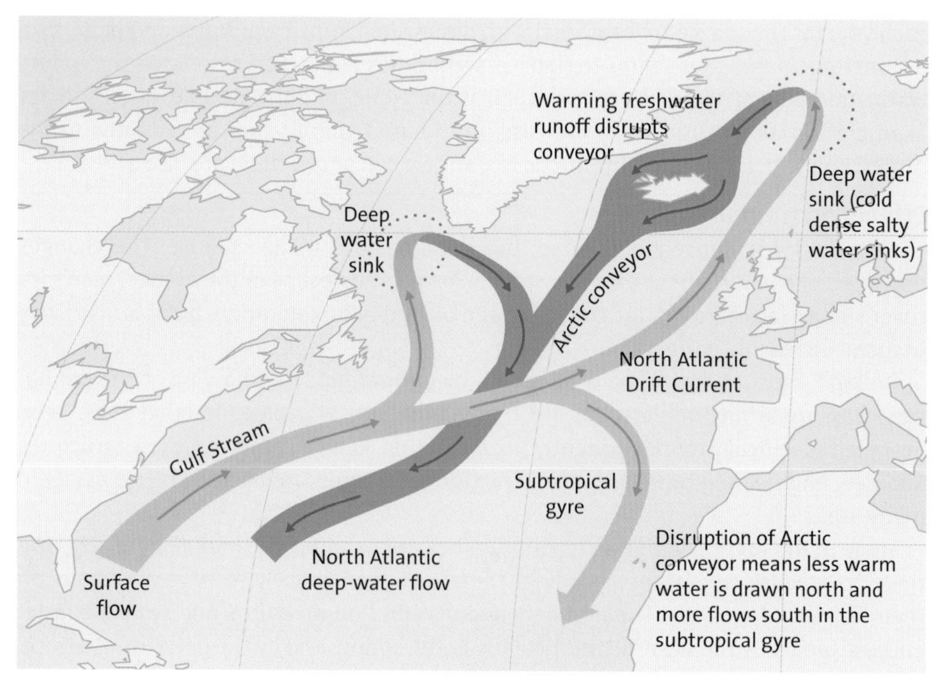

Figure 4.3 *Effect of melting Arctic ice on ocean currents*

standards of living may rise. The increase in the number of herring and cod migrating and breeding in the waters may also increase opportunities for commercial fishing.

Other possible impacts

A great deal of carbon is stored in the Arctic permafrost. As the permafrost continues to thaw, some of this carbon will be released into the atmosphere as methane. It is thought that this will add to the greenhouse gases and add to global warming. On the other hand, increased temperatures will result in longer growing seasons and the northward movement of vegetation could increase the capture and storage of carbon by photosynthesis.

Some scientists also believe that the melting of ice in the Arctic could make the Arctic Ocean less saline and warmer. This would weaken the formation of the Arctic Conveyor (also known as the Atlantic thermohaline circulation), which draws the warm Gulf Stream current northwards. The loss of the warm Gulf Stream would cause dramatic cooling of the climate in northwest Europe — the UK would have a climate similar to that of Siberia (Figure 4.3).

Using case studies 5

Question

Assess the impacts of climate change on Arctic regions.

Guidance

Being brief and direct, this seems a straightforward question. However, good examination technique should enable you to recognise that the question is requiring a discussion of 'impacts' — plural. This indicates that you should focus on a *range* of impacts. Case study 5 refers to many possible impacts, including those on the physical environment, on the natural world and on human activities. Once again, the recognition that a range of impacts needs to be discussed provides evidence of synopticity. Make sure that you are precise in the description of these impacts and, where possible, support your answer with references to places.

The second element to notice in this question is the overall command 'assess'. Assess asks for a statement of the overall extent, or importance, of the feature or issue being considered. You are being asked to weigh up the overall effect of an issue, or set of issues, and come to a conclusion. In this case, you could summarise by saying which of the impacts, if any, are likely to have the most long-term impact. The command is open-ended — any conclusion is valid, so long as it is based on solid evidence that has been presented.

THE ANTARCTIC AND GLOBAL WARMING

Case study 6

The Antarctic ice sheets form the largest mass of ice on the planet. They cover a continent that is more than twice the area of the USA and in places they are over 4 km thick. This constitutes an enormous volume of ice — enough to raise global sea level by 65 m if it were to melt.

The ice sheets

There are two ice sheets in Antarctica, with very different characteristics. The largest, the East Antarctic Ice Sheet (EAIS), covers most of the continent. The bedrock on which it sits lies above sea level in most places, and so it is termed a terrestrial ice sheet.

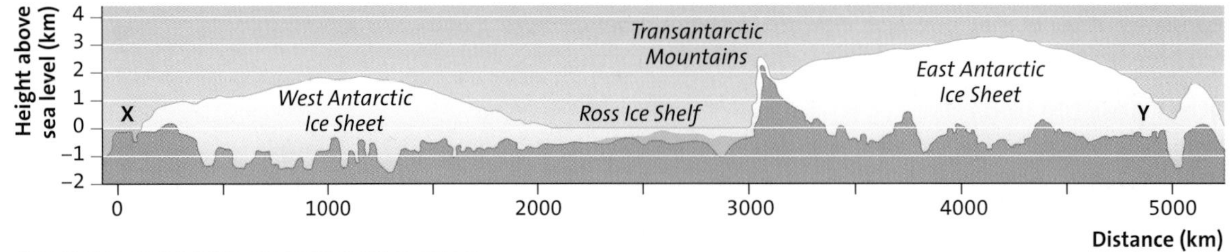

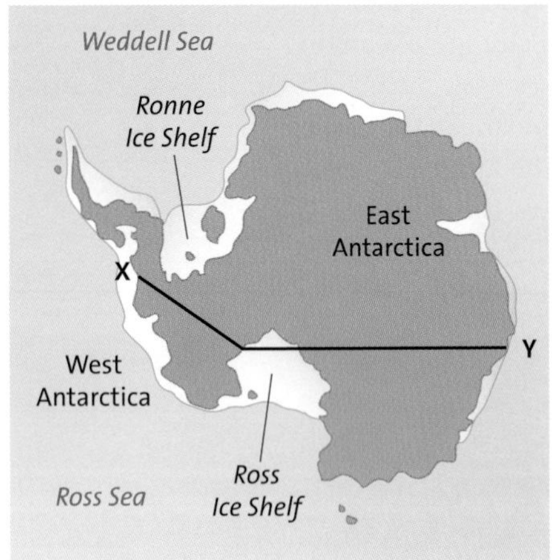

Figure 4.4 *Cross-section through Antarctica showing the East and West Antarctic Ice Sheets*

In contrast, the smaller West Antarctica Ice Sheet (WAIS) is a marine ice sheet, because it has a bed that is below sea level over most of its area. The difference in bedrock elevation may seem minor, but it is in fact crucial because it is thought to determine the relative stability of the two ice sheets (see Figure 4.4).

The EAIS

The EAIS is thought to be relatively stable. Scientists have shown that during a natural warm period 2–5 million years ago the EAIS did not reduce significantly in size. It is so cold in Antarctica that there is virtually no surface melting of the ice. Even if the temperature increased by a few degrees it would still be far too cold for surface melting and so the ice sheet would not shrink. Only if the temperature went up by huge amounts (tens of degrees) would it be possible for major melting to begin.

The amount of snowfall on Antarctica also depends on temperature. The warmer the air, the more moisture and snow it can hold. This means that as temperatures rise from the current very cold averages, the amount of snowfall on Antarctica will *increase* and the EAIS may even get *bigger*. However, as temperatures rise further this effect will be outweighed eventually by the onset of melting.

The conclusion is that in the next few decades, and perhaps even centuries, the increased snowfall on the EAIS may extract water from the oceans, therefore partly offsetting sea-level rise from other sources — for example, melting of small ice caps (such as in Alaska) and thermal expansion of the oceans. Most scientists working on the Antarctic ice sheets think that the EAIS will not collapse or cause a significant sea-level rise for many centuries to come.

The WAIS

Marine ice sheets such as the WAIS are thought to be inherently unstable. The fact that much of the ice sits below sea level means that they are sensitive to small rises in sea level, which can cause them to thin. Moreover, the WAIS is drained by several ice streams — fast-moving 'rivers' of ice very different from the slow-moving ice of the rest of the Antarctic ice sheets. Because they move so fast, and drain so much of the ice in the WAIS, the ice streams have the potential to rapidly increase the amount of ice being lost from the ice sheet to the ocean.

As the WAIS alone has the potential to raise sea level by 5 m, scientists have been trying for decades to understand the ice sheet. First, they have discovered that in the past some ice streams have speeded up and slowed down by large amounts. This is important because it shows that the speed of ice streams can vary to such an extent that the amount of ice being drained from the WAIS may change. Indeed, one ice stream stopped completely about 200 years ago.

A second discovery is that the WAIS may have collapsed several hundred thousand years ago during a warmer period. The evidence for this is the discovery of small fossil sea creatures below a currently active ice stream, along with high concentrations of the isotope beryllium-10, which could only accumulate if the ice stream bed had been exposed to the atmosphere. This suggests that at some point the ice stream must have been absent (along with all the ice it was draining), exposing the bed to the atmosphere and allowing the sea to wash in and deposit the fossils.

What could have caused this collapse? Scientists believe that two processes may operate to cause this. The first is the loss of fringing ice shelves, which allows more of the land-based ice to move into the ocean. The removal of ice shelves may also allow warm ocean currents to reach the base of the ice streams. The second process is the speeding-up of the large ice streams. This has been measured recently in an area called the Amundsen Sea Embayment. Here, two or three ice streams have accelerated and are discharging greater amounts of ice into the ocean than before. The reasons for this acceleration and the extent to which it may penetrate inland are now being researched. It is not known to what extent climate change may push the WAIS onto a slippery slope towards future collapse.

The Antarctic Peninsula

The Antarctic Peninsula, the finger of the continent pointing towards South America, has experienced rapid warming in recent decades (Figure 4.5). There have been a number of impacts of climate change, ranging from physical effects such as the collapse of ice shelves and the retreat of glaciers, to biological effects such as changes to vegetation and movements in penguin colonies.

Temperature records show pronounced warming here in recent decades. Readings have been taken at a number of sites, including Vernadsky station in the western part of the peninsula. This station is a Ukrainian research base, which was once a British base (Faraday base). Since 1947, records have been kept by both sets of scientists. The temperature record from this station shows a warming of 2°C since the 1950s with the warming trend strongest in the winter months. In addition, temperature measurements in the troposphere above the surface of Antarctica also show a warming during the winter of 0.5–0.7°C per decade over the last 30 years. This is the largest tropospheric warming anywhere on Earth. In the ocean to the west of the Antarctic Peninsula, monitoring has detected a 1°C warming since the 1950s.

Climatologists suggest that this warming is due to increased atmospheric circulation caused by the effects of global warming, which has brought more warm moist air from further north down towards the peninsula. In the western part of the peninsula there is a strong correlation between sea-ice extent and winter air temperatures — when sea ice is less extensive, the winters are much warmer and vice versa. This is because the sea ice acts as a cold lid on the ocean. When there is less sea ice, more heat can escape from the ocean (which is about 0°C) and warm the atmosphere (about –10°C). The sea warms the air, which in turn may cause a reduction in sea ice and so on.

What are the impacts?

The most pronounced impact has been the collapse of some Antarctic Peninsula ice shelves. The warmth has caused extra melting on the surface of the ice shelves, which

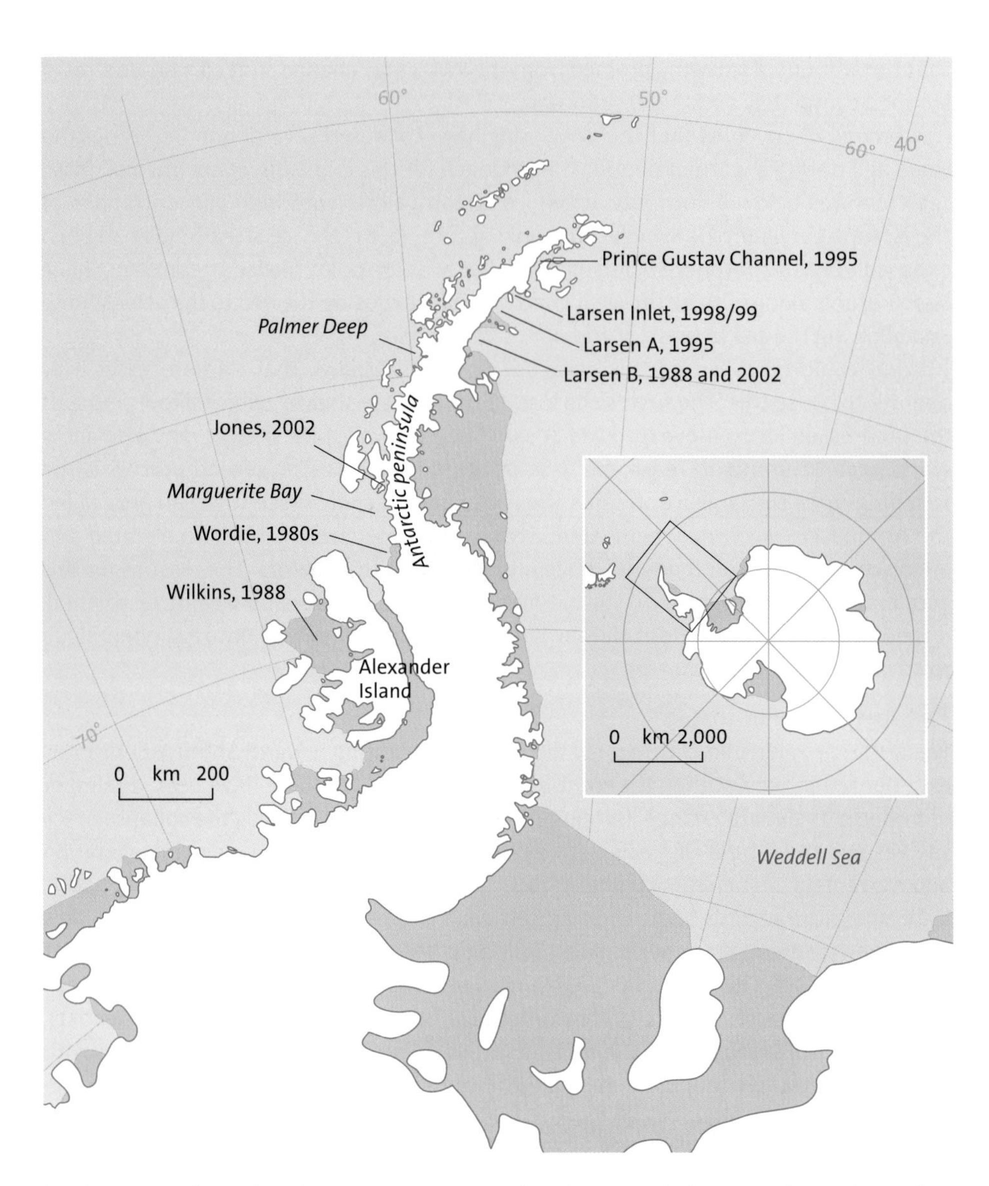

Figure 4.5
The location of the Antarctic Peninsula and ice shelves that have collapsed

leads eventually to break-up. Recent research using aerial photographs and satellite imagery has shown that nearly 90% of the glaciers in the peninsula have retreated since they were first measured. Moreover, there has been melting of the surface of the ice sheet covering the peninsula.

There have also been impacts on the wildlife of the peninsula. There are only two flowering plants in Antarctica — Antarctic pearlwort and Antarctic hair grass. Both have shown recent increases in species density and range (they are appearing further south). Some lakes on Signy Island, to the northeast of the peninsula, have been monitored for decades and are showing changes. For example, the number of ice-free days during summer has increased by over 30 days since 1980. The temperature of the lakes has risen by 1°C and the biological productivity has increased four-fold in the same period.

Different penguins have different tolerances for sea ice. Some, such as the adelie penguin, favour some winter sea ice. Others, for example the chinstrap (Figure 4.6) and

Figure 4.6
Chinstrap penguins will only breed where there is little sea ice

gentoo penguins, are less tolerant and will only breed where there is little or no sea ice. Adelie penguins seem to be moving south; the chinstrap and gentoo penguins are now more common in areas dominated formerly by adelies.

Marine organisms are not immune from the effects of warming. Invertebrates such as limpets and clams that are adapted to the cold waters around Antarctica have a limited tolerance of the warming ocean. The warmth does not necessarily kill them immediately, but they lose the ability to carry out simple actions such as swimming (scallops) or turning over when knocked (limpets). In some cases, this occurs after only 2–3°C warming and it can cause them to become less effective in feeding or avoiding predators.

Using case studies 6

Question

Compare the effects of climate change on different parts of Antarctica.

Guidance

Antarctica is a vast continent, almost the size of North America, yet many writers in the media refer to it as a complete, and single, entity when climate change is being discussed. There are several differences regarding the impact of climate change. The EAIS, which covers by far the largest area, seems to be largely unaffected by climate change. On the other hand, the smaller WAIS is seemingly hugely affected. The one significant area of land that is not covered by an ice sheet — the Antarctic peninsula — is demonstrating many of the effects shown at the Arctic. There are manifestations of impacts both on the physical landscape and on the natural world.

Note also the command word to this question: compare. You should provide a point-by-point account of the similarities and differences between the various areas referred to (as suggested above). Note that two or three separate accounts without any links do not comprise a comparison — you should not expect the examiner to do that on your behalf. A good technique would be to use comparative adjectives, or adverbs, or even to use link words such as But..., However..., On the other hand..., Yet..., Another aspect to consider is..., In contrast..., Conversely...
Another option would be to draw an annotated sketch map.

Case study 7 PERMAFROST AND GLOBAL WARMING

Permafrost is ground in which the temperature is below freezing for more than 2 years. It occurs in a wide area around the circumference of the Arctic, both on land and in places on the sea bed (Figure 4.7). In the coldest part of the Arctic the permafrost is hundreds of metres thick, whereas towards the southern fringe there may be only a few metres of frozen soil (Figure 4.8). In the southern parts of the permafrost zone, the soil thaws a little on the surface during summer, perhaps to a depth of about a metre. This layer of summer melting is called the **active layer**, but the remainder of the permafrost below this stays frozen year after year, and in places has been frozen for millions of years.

Twenty to twenty-five per cent of the world's land area is underlain by permafrost, a figure that rises to 50% in countries such as Canada and Russia. In these countries, permafrost causes considerable problems for economic activities:

- In construction:
  - the elasticity of ice reduces the effectiveness of explosives used in excavating
  - structures have to be insulated from the ground so that the soil does not melt and sag or move under them
  - paved roads and runways must have insulation beneath them
- Normal, agricultural activities are impossible.
- Funerals are affected because it is impossible to dig graves in the winter.

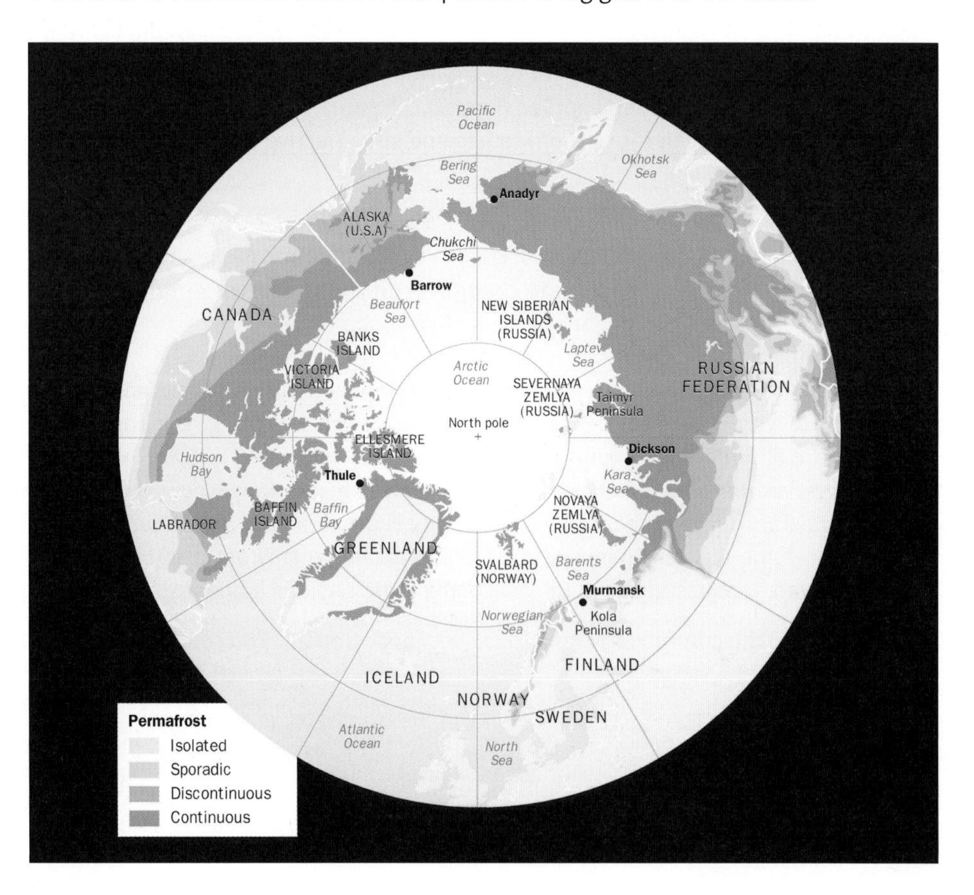

Figure 4.7 *Permafrost distribution in the Arctic*

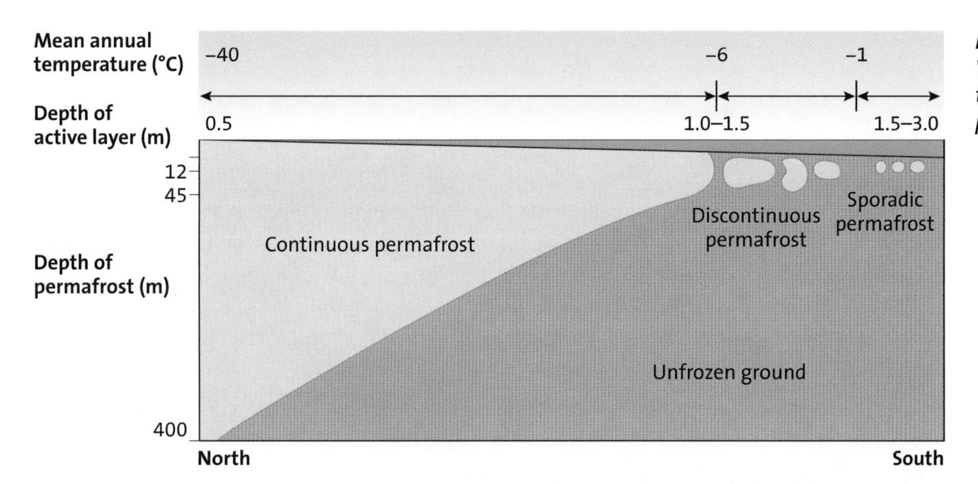

Figure 4.8 *Variations in the depth of permafrost*

Scientists have already measured a retreat in the permafrost zone as climate warming has increased. This is particularly acute in northwestern Canada and in Siberia. Increases in temperature of only 1°C have led to a trebling of thaw rate in parts of central Canada. The immediate impacts of melting can be seen in the Arctic communities — buildings become undermined, roads subside unevenly and crack, the supports holding pipelines can shift and even crack the pipeline. Given the world's dependence on oil and gas, the threat of pipelines having to be shut down is alarming. On a smaller scale, it is common to see trees leaning over as their roots are no longer supported firmly in the frozen ground.

Another major concern regarding the melting of permafrost is the release of organic carbon. The soils of the permafrost are normally crammed with undegraded, well-preserved organic matter in the form of leaves, twigs, roots etc. This is an enormous store of carbon, kept inert by being frozen in the ground. However, if the ground were to melt and the organic matter start to rot, this carbon would be released as either carbon dioxide or methane — more greenhouse gases. This would melt more permafrost and so on, in a worsening positive feedback cycle. The Arctic is estimated to contain about 900 gigatonnes (Gt) of carbon. Humans emit about 9 Gt of carbon from fossil fuels and deforestation every year. Hence, it would only take the release of 1% of carbon from Arctic permafrost soils in a year to effectively double the emission of greenhouse gases, especially methane.

Scientists are now starting to put together monitoring networks to measure the release of carbon from Arctic soils. They have shown that carbon is released following warming and have also discovered two other effects. First, the rate of soil thawing and carbon release seems to slow down after the initial burst. This could be because, as the soils warm up, new plants such as mosses grow on the surface and insulate the soil from further warming. In addition, the warmer climate means less snow in winter, which allows the winter cold to penetrate further into the soil, in effect storing up 'cold' in the soil to protect it from the warmth of the following summer.

On the other hand, the melting of the permafrost releases a lot of water onto the ground surface, creating a series of thaw lakes that are unable to drain because of the remaining frozen and impermeable soils beneath them (Figure 4.9). The organic matter that starts to decay does so under water, and therefore in the absence of oxygen. This means that the carbon is emitted in methane rather than in carbon dioxide. Methane is

Figure 4.9
An aerial view of summer tundra with ponds in western Siberia

Bryan and Cherry Alexander

a more potent greenhouse gas than carbon dioxide, so the overall effect of the thawing is increased by the thaw ponds.

Scientists have predicted that the projected loss of permafrost by 2100 could range from 60% to 90%. Although permafrost may seem an irrelevance to those of us who live in the temperate latitudes, it is clear that changes in it may have significant consequences for everyone through the impact of greenhouse gases.

7 *Using case studies*

Question

'Buildings constructed on permafrost are becoming increasingly unstable, requiring expensive repair. New buildings will require different, and likely more expensive, construction techniques. Roads and airport runways built on permafrost are now unstable and hard to maintain. The...permafrost and sea ice are all critical to the geography and economy of Greenland as a whole, and essential to the economy, social structure and culture of Inuit people.'

Extract from *Climate Change in Greenland: Impacts and Response*, by Anna Heilmann

Outline how the impact of climate change on the permafrost is having a significant effect on the people of Greenland.

Guidance

Here is an opportunity for you to investigate and consider the impact of climate change on a particular group of people. The extract in the question is taken from a brief report by a member of the Inuit community in Greenland. Whereas the case study provides an overview of the problems that the melting of the permafrost causes, you are being asked here to consider the impact on a specific area and on the people who live there.

Clues are present in the quotation:

- economy: to do with money, work, industry, jobs and prospects
- social structure: to do with people, their quality of life, health, education and prosperity
- culture: their traditional way of life, social norms and the aspects of their existence that make them distinctive

THE BRITISH ISLES AND GLOBAL WARMING

As a result of global warming, the UK could experience warmer summers, longer hot spells, droughts and increased storm activity. In contrast, some climatologists believe that changes in the pattern of ocean currents could result in much colder conditions due to the weakening of the Arctic conveyor (see p. 23).

In coastal areas, increases in mean sea levels, the frequency and magnitude of storms, storm surges and waves would all lead to more coastal flooding. Sea levels around Britain are predicted to rise by between 12 cm and 37 cm by 2050, which would make a number of low-lying areas vulnerable, particularly the coasts of East Anglia, Lancashire, the Humber estuary, the Essex mudflats, the Thames estuary, parts of the north Wales coast, the Clyde/Forth estuaries and Belfast Lough. Flooding would lead to disruption in transport, manufacturing and the housing sector. In addition, there would be longer-term damage to agricultural land and coastal power stations, and water supplies could be contaminated by salt infiltration.

Climate changes are likely to have a substantial effect on plant growth and, by extension, plant productivity in agriculture. Higher temperatures could result in:

- a decrease in yields of cereal crops
- an increase in yields of sugar beet and potatoes
- an increase in the length of the growing season for grasses and trees, bringing higher productivity
- the introduction of new crops and species — the UK could become a major wine-producing region, with rice cultivation also being a possibility in the Fens (see below)
- an increase in some pests, such as the Colorado beetle, which causes serious damage to potatoes

In terms of flora, fauna and the landscape, a sustained rise in temperature could have the following effects:

- a significant movement of species northwards and to higher elevations
- the extinction of native species that are unable to adapt to the increasing temperature, e.g. subarctic plants in the Cairngorms
- the loss of species that occur in isolated damp, cool or coastal habitats
- the invasion and spread of alien weeds, pests, diseases and viruses
- an increased number of foreign species of invertebrates, birds and mammals which may out-compete native species
- the disappearance of snow from the tops of the highest mountains

For soils, higher temperatures could reduce the water-holding capacity, increasing the likelihood of soil-moisture deficits. The stability of building foundations and other structures, especially in central, eastern and southern England, where clay soils with large shrink and swell potential are abundant, would be affected if summers became drier and winters wetter. There could be a loss of organic matter, which would affect the stability of certain soil structures. Soil structure could also be affected if the water table rose with rising sea levels.

Water resources would benefit from wetter winters, but warmer summers with increased evaporation could have the opposite effect. In terms of energy resources, higher temperatures could decrease the need for heating, but a growing demand for air conditioning would increase electricity consumption.

In July 2009, the Department for Environment, Food and Rural Affairs (**DEFRA**) produced a report giving an assessment of future possibilities for British food production by 2030, based on recent climate data. The forecasts highlight some of the unexpected benefits of a warmer climate:

- The south coast will provide an attractive climate for the cultivation of olive trees, apricots and garlic.
- Dates could be grown in Devon and Cornwall.
- Other crops that could be possible include chick peas, figs, aubergines, peppers and chillies, all on a commercial scale.
- Lavender, already being cultivated in small amounts, could become more widespread.
- Vineyards cultivating grapes such as pinot noir (champagne) and tempranillo (rioja) will be common.

However, one of the major problems will still remain — the management of water supplies for farming. By 2050, the Environment Agency expects the supply of water to fall by 15%.

Using case studies 8

Question

Carry out a cost–benefit analysis of the effects of climate change on the UK.

Guidance

A cost–benefit analysis (CBA) assesses the desirability of a project, or an event. In the case of climate change in the UK, the proposed outcomes are not entirely negative. There are several positive outcomes as well. A CBA evaluates and tabulates the social and economic costs of an event against the social and economic benefits it will bring. A suggestion therefore is to draw up a matrix with three columns, and a series of rows such as:

Impacts	Costs	Benefits
Impact on landscapes		
Impact on agriculture		
Impact on flora and fauna		
Impact on structures		
Impact on water supply		
Impact on...		

Case study 9 THE AFRICAN SAVANNA AND GLOBAL WARMING

The tropical wet–dry savanna climate is experienced over a huge area of Africa (Figure 4.10). It surrounds the rainforests that hug the equator on the western side of the continent, to both the north and south. It is essentially a zone of transition close to the equator where the dry season is very short; this dry season increases in duration with an increase in latitude. Opinions are divided when it comes to assessing the likely impacts of climate change within this region and a significant amount of uncertainty is apparent. However, many scientists believe the following:

- Overall, the savanna lands are likely to experience an increase in temperature of some 1.5°C by 2050. Surrounding sea temperatures are not expected to rise to the same extent (0.6°C–0.8°C) resulting in a greater temperature differential between land and sea.
- Precipitation is expected to show an overall increase of some 15% within the savanna lands closer to the equator, but in areas towards the northern and southern fringes of the climate zone it might actually decrease by 10%, for example in the Horn of Africa. This is likely to be due in part to the Sahara heating up more than the Atlantic Ocean, causing more moisture to be drawn in from the ocean during the wet season. So, rainfall amounts could go up and down in the same climate zone.
- An increase in the variability of rainfall will lead to more frequent droughts and flooding in some areas. During the wet season, 25–50% more rainfall is expected to fall. More frequent droughts will contribute to desertification, particularly along the extreme fringes of the savanna biome. However, in areas where annual rainfall remains below 650 mm the savanna ecosystem is likely to be more stable than in those regions where rainfall totals rise above 650 mm. Consequently, pastoral nomadic farmers, who do not rely quite as strongly on reliable rainfall as those subsistence farmers who depend on rain-fed agriculture, are less likely to be affected by climate change in these areas.

Figure 4.10 *The tropical wet–dry regions of Africa*

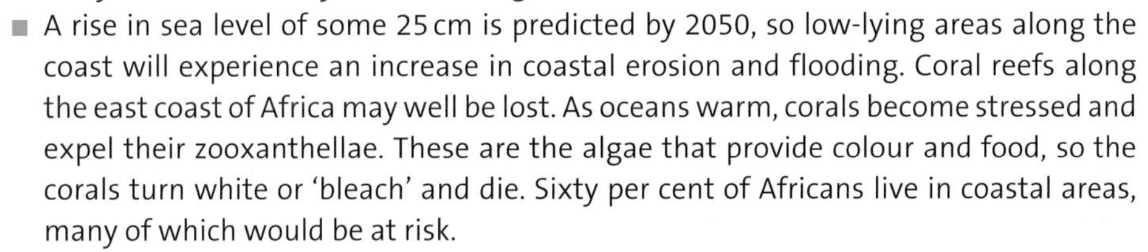

- A rise in sea level of some 25 cm is predicted by 2050, so low-lying areas along the coast will experience an increase in coastal erosion and flooding. Coral reefs along the east coast of Africa may well be lost. As oceans warm, corals become stressed and expel their zooxanthellae. These are the algae that provide colour and food, so the corals turn white or 'bleach' and die. Sixty per cent of Africans live in coastal areas, many of which would be at risk.
- Higher rainfall totals in savanna lands closer to the equator are likely to result in an increased growth of trees and scrub in what was previously predominantly grassland. Changes to the tree–grass balance will have an impact on plant and animal life and on the water and carbon cycles. Increased rates of evapotranspiration may result in lower annual discharge in rivers such as the Nile and the Zambezi.
- Vector-borne diseases (e.g. malaria) and water-borne infections (e.g. diarrhoea) could increase with climate change as there will be more water lying undrained.

THE AMAZON BASIN AND GLOBAL WARMING

Case study 10

The Amazon basin in South America, vegetated by tropical rainforest, lies within the equatorial climate zone and covers an area of some 8 200 000 km², mainly in Brazil. The Amazon River flows through the basin from its source high in the Andes Mountains towards its mouth in the Atlantic Ocean (Figure 4.11). It is the largest single source of freshwater runoff on Earth, representing 15–20% of global river discharge. At present,

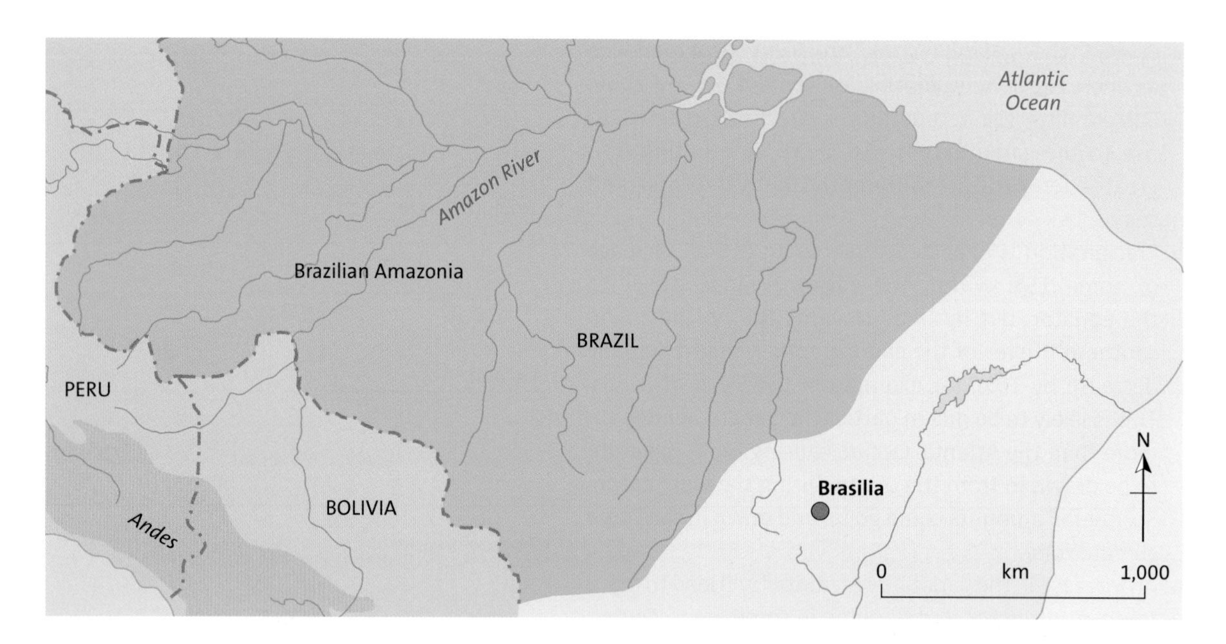

Figure 4.11 *The location of the Amazon River*

the Amazon rainforest acts as a carbon sink and absorbs around 35% of the world's annual carbon dioxide emissions and produces more than 20% of the world's oxygen. The Amazon rainforest contains the greatest biodiversity on Earth, providing a habitat for more than half of the world's estimated 10 million species of plants, animals and insects. Predicted impacts of climate change on this region include the following:

- There could be an increase in temperature of 2–3°C by 2050, which is likely to result in increased rates of evapotranspiration and a more vigorous hydrological cycle. Sea temperatures are expected to warm too, particularly in the Pacific Ocean. This may have a knock-on effect on the ENSO, which is likely to occur more frequently (p. 16).
- A decrease in precipitation during the dry season (which can last up to 4 months of the year) is expected. Reduced rainfall and prolonged drought are features of an El Niño year and these could, therefore, be experienced more frequently. It is also possible that there will be more intense rainfall during the wet season.
- Sea levels are rising currently by some 5 mm y-1 along the delta of the Amazon. Increased erosion and flooding is likely to have a substantial impact on low-lying areas and could destroy the coastal mangrove forests.
- Up to 40% of plant species may become unviable in the Amazon rainforest by 2080. Large areas of the evergreen tropical rainforest may be succeeded by mixed forest and savanna grassland vegetation.
- As the dry season lengthens, trees will have more time to dry out, so there is likely to be an increased incidence of spontaneous forest fires. This would add to carbon dioxide emissions.
- By 2050, forest die-back as a result of vegetation succession and fire is predicted to result in the Amazon region becoming a net source of carbon dioxide, rather than a carbon sink, exacerbating the rate of global warming.
- Glaciers in the Andes provide the source for as much as 50% of the discharge of the Upper Amazon. Over the last 30 years, Peruvian glaciers have shrunk by 20% and it is predicted that Peru will lose all its glaciers below a height of 6000 m by 2050. This will have a further impact on the Amazon's hydrological cycle.

BANGLADESH AND GLOBAL WARMING

Bangladesh lies within the global climatic zone that experiences the tropical wet monsoon climate (Figure 4.12). With a population estimated by the United Nations to have been in excess of 150 million in 2008, it is one of the most densely populated countries in the world, with some 1102 people per km². As one of the poorest nations in the world it is also one of the most ill prepared to face the challenges presented by climate change and is likely to be one of the worst affected by global warming. Predicted effects include the following:

- As a result of warmer sea temperatures in the Bay of Bengal, there will be 10–15% more precipitation annually by 2050 and an increase in the frequency and severity of cyclones during the wet season.
- There will be a 20% increase in river discharge, due in part to the predicted increase in precipitation but also as a result of glacier melt in the Himalayas, where the rivers Brahmaputra, Meghna and Ganges have their sources.
- There will be a significant rise in sea level along the coastline and inland along the countless tidal inlets. In 2001, the World Bank reported rising sea levels of some 3 mm y-1 (compared with the world average of 2 mm) and predicted that by 2050 a 1 m rise in sea level is possible if no preventative action is taken. This would result in some 15% of the total land area of Bangladesh being inundated by salt water, as much of the land is close to sea level.

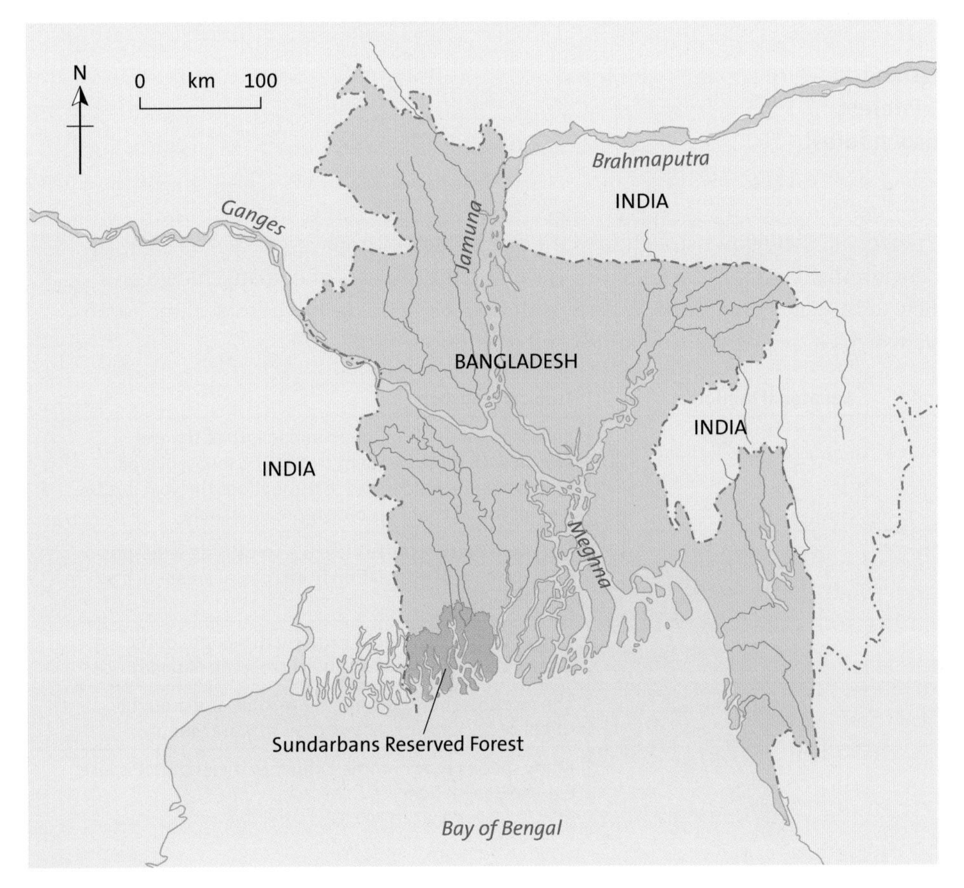

***Figure 4.12** The location of Bangladesh and the Sundarbans*

- An estimated 13–30 million people could be displaced from their homes by permanent flooding and the total annual crop of rice is likely to fall by at least 30% because of the loss of paddy fields. In a report published by the World Bank and the United Nations Climate Change Secretariat in 2001, the conclusion was that: 'Bangladesh is used to coping, but rising sea levels pose new dangers.' Loss of land on such a large scale is likely to result in mass migration into northeast India and there is likely to be increased international tension between the two countries, in addition to internal political instability.
- Coral reefs will become irreparably damaged by severe storms and this will have a knock-on effect on rare marine species, such as dolphins and turtles.
- The Sundarbans is a cluster of islands along the coast, with a total area of about 10 000 km². It stretches from Bangladesh into India. It is home to the world's largest natural mangrove forest and was declared a UNESCO World Heritage Site in 1997. The Sundarbans is Asia's largest natural 'carbon sink' and is the home to many species, including the endangered royal Bengal tiger, the Indian python and the estuarine crocodile. Climate change is likely to be partially responsible for the loss of this unique area and will lead to complete flooding of low-lying 'char' islands, retreat of shorelines, salinisation of the soil and a rise in the water table.

Question

Tabulate a summary of the effects of climate change on each of the following tropical regions:

- **savanna grassland**
- **equatorial rainforest**
- **tropical monsoon forest**

Guidance

This is a simple technique that allows straightforward comparison of impacts. Case studies 9, 10 and 11 refer to environments in tropical latitudes. However, the impacts of climate change on them are not likely to be exactly the same. A table such as the one below (with tropical monsoon completed for you) helps in seeing the similarities and differences easily.

Savanna grassland	Equatorial rainforest	Tropical monsoon
		Increased total precipitation and length of the wet season with more frequent tropical cyclones; overall temperatures could increase throughout the year but the wet monsoon could become more unreliable
		Increased flooding as sea levels rise, causing widespread displacement of population from coastal areas such as Bangladesh
		Longer wet season might increase the length of the growing season and allow multi-cropping of paddy rice
		The monsoon may become less reliable and drought might occur, causing a huge crisis in water supply
		Many species may become extinct as their habitat is lost, e.g. the Bengal tiger

TROPICAL LOW-LYING ISLANDS AND GLOBAL WARMING

Case study 12

Low-lying islands are already experiencing the effects of sea-level change. Islands such as those in the Pacific have common vulnerabilities (Figure 4.13):

- small physical size — there is nowhere for people to move or evacuate to
- low-height land and rising sea level
- prone to the effects of tropical storms, which raise the sea level
- dense and growing populations, with some experiencing rapid urbanisation
- increasing degradation of the coral reefs and atolls that make up the islands
- contamination of groundwater supplies by sea water
- limited resources other than the sea and tourism

There are more specific issues:

- In the Maldives over 300 000 people live on over 1100 islands, most of which lie only 2 m above sea level. However, it is the loss to the rising sea of the thin layer of fresh groundwater, on several islands only 10 cm deep, that is the biggest and most immediate threat.
- On Carteret Island (the Pacific) 2000 people have been forced to leave the island due to water supplies becoming more saline.
- Most of the Marshall Islands inhabitants (57 000) live on land at 1 m above sea level. In January 2009, high tides and strong winds caused sea water to flood parts of the low-lying areas in the capital Majuro.
- In Kiribati, two islands have already disappeared because of rising sea levels.
- In Vanuatu, the island of Tegua, north of the main island of Efate, was almost abandoned in December 2005 and is likely to be the first community in the world to be forced out by rising sea levels (climate change refugees?).

Figure 4.13 *Pacific islands will be the first to suffer from sea-level rise*

Corel

- It is thought that the whole island of Tuvalu (11 000 people) will have to be evacuated by the end of the century. Some have already moved to New Zealand; others may have to go to Fiji.

Tuvalu

Tuvalu lies in the Pacific Ocean between Australia and Hawaii. The population of 11 000 is spread over nine islands, none of which is over 9 m above sea level. As Tuvalu is too remote for large-scale tourism, its economy relies on farming, fishing and foreign aid.

As sea levels rise by 1–2 mm per year the low-lying islands are being flooded more frequently by the highest tides and tropical storms. The porous coral atolls flood from below as water levels rise, pushing salt water up through the ground in numerous little springs. This poisons farming areas, including the pits in which pulaka tubers are grown — an important food crop for the locals. Poor coastal management and beach mining for building materials worsen the problem and road building projects have stripped valuable land of vegetation. If sea levels rise by 20–50 cm by the end of the century, as the IPCC predicts, Tuvalu may soon become uninhabitable.

Question

How are small island developing states (SIDS) such as those in the Pacific threatened by climate change?

Guidance

SIDS are particularly vulnerable to climate change. It will probably cause:

- higher temperatures
- less reliable rainfall
- more tropical storms
- stronger tropical storms
- rising sea level
- changes to ecosystems (including coral reefs)

More and stronger tropical storms can cause flooding, leading to:

- landslides on deforested slopes
- salt water inundation
- destruction of homes, infrastructure, farms, tourism etc.

Rising sea level can lead to:

- erosion of shoreline and protective barrier reefs
- flooding of land and settlements
- changing ecosystems in coral reefs, often causing destruction
- pollution of fresh water
- migration away from the islands

Warming of climate can cause:

- deaths, particularly of old people and babies
- invasion by new diseases adapted to the warmer climates; these can affect agriculture as well as people
- loss of ecosystems that may attract tourists (e.g. in the Maldives)

Part 5

Organisational statements on climate change

A large number of supragovernmental, governmental and non-governmental organisations (NGOs) around the world have established themselves, or been established, to investigate climate change and to make recommendations as to how the world should respond to it.

At a global level, by far the most influential is the Intergovernmental Panel on Climate Change (IPCC). It has produced four Assessment Reports, each updating its predecessor. Within the UK, the *Stern Review* was commissioned by the government to examine the phenomenon and to make recommendations. In addition, The Tyndall Centre for Climate Change Research and the Royal Society are organisations based in the UK that bring together scientists, economists, engineers and social scientists to comment on climate change (see websites on page x).

Elsewhere, 'green' NGOs such as Greenpeace have made their views known; even supranational bodies aimed at other aspects of human life, such as the World Health Organization (**WHO**), have made comments on how they see the impact of climate change.

This section examines the reports of a selected few of these organisations.

The Intergovernmental Panel on Climate Change (IPCC)

The IPCC was formed jointly in 1988 by two United Nations bodies: the World Meteorological Organization and the United Nations Environment Programme. Its mandate was to prepare authoritative assessments of human-induced climate change covering the fundamental science, the likely impacts on human communities and ecosystems, and the policy options including economic, technical, social and political aspects.

In the preparation of its reports, the IPCC has involved a wide range of scientists from many countries, including a number of the world's leading climate scientists. Wide ownership of its reports has therefore been achieved by the international scientific community. To ensure accuracy and quality, the reports have been subjected

Table 5.1 *The increase in contributions to IPCC reports from scientists and governments over time*

	1990 First Assessment Report (FAR)	2007 Fourth Assessment Report (AR4)
Pages in report	365	987
Authors (lead and contributing)	170	552
Reviewers	200	600
Number of countries attending plenary session	35	113

to extensive review procedures. The involvement of government representatives in the preparation of the summaries for policy-makers that accompany each report has ensured their relevance to policy formation and has also meant that governments have felt ownership of the reports. The degree to which contributions from scientists and governments have increased over time is shown in Table 5.1.

The IPCC's first assessment report (FAR) was prepared in 1990. The achievement of such a broad scientific consensus provided a firm basis for agreement by the world's governments to the United Nations Framework Convention on Climate Change (**UNFCCC**) at the Earth Summit in Rio de Janeiro in 1992. Subsequent comprehensive IPCC reports have been published in 1995 (SAR), 2001 (TAR) and 2007 (AR4).

The IPCC has a task force on national greenhouse gas inventories and three working groups:

- Working Group I assesses the scientific aspects of the climate system and climate change.
- Working Group II assesses the vulnerability of socioeconomic and natural systems to climate change, negative and positive consequences of climate change, and options for adapting to it.
- Working Group III assesses options for limiting greenhouse gas emissions and otherwise mitigating climate change.

Predicting future emissions and their impacts

How climate will change during the twenty-first century is a critical question that the IPCC and climate change scientists are addressing. Simulations are used to create models of climate change. Predictions rely on three main scenarios — low, medium and high greenhouse gas emissions — which produce different global warming outcomes. Higher emissions are thought to lead to higher temperatures.

Predicting future climate change requires a sequence of steps, each of which has associated uncertainties. First, emissions of greenhouse gases and aerosols have to be identified specifically. Their dependence on unknown socioeconomic behaviour is modelled by using a range of scenarios designed to suggest what may occur. In reality, feedbacks are likely to occur — for example, climate change is likely to influence socioeconomic behaviour.

The next step is to determine what the impact of these emissions will be on the composition of the atmosphere. While there is no doubt that increasing greenhouse gas emissions will cause adverse impacts, the range of results from climate models makes it difficult to predict just how serious these will be. For example, some scientists are investigating the potential for rapid methane release from the melted permafrost. It has also been suggested that changes in land cover, such as large-scale planting of forests, may have an effect on climate by altering the reflective property of the Earth's surface (the albedo).

There is agreement that the latest climate models are able to predict the long-term temperature record more accurately than previously. However, uncertainties in results from climate models remain significant and arise from both the models themselves and from the choice of scenarios used.

To describe climate change in terms of changes to global averages provides limited information (best estimates are that temperatures will rise on average between 1.8°C and 4°C by the end of the century (Figure 5.1)). Most of the impacts of climate change depend on changes in regional climate or on changes in the natural variability of climate, including extreme weather events. Will we see changes in regional climate patterns such as the El Niño or the Indian monsoon? Will there be an increase in extreme winds in Europe? To answer questions like these, we will have to improve our ability to predict changes in weather phenomena. To do this, we will need to use a large amount of computing power to perform calculations using improved and higher resolution climate models. The models will need to use

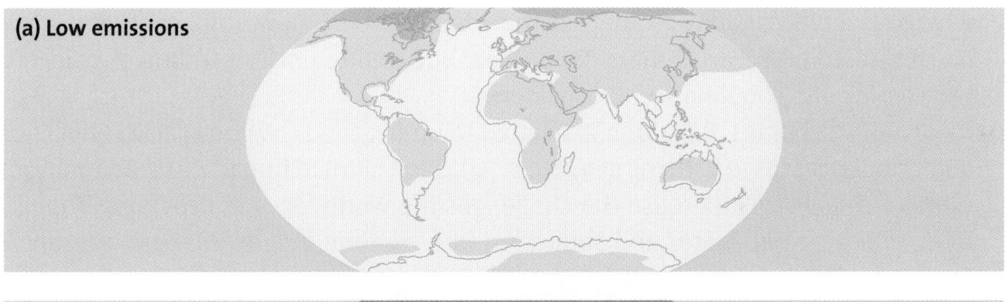

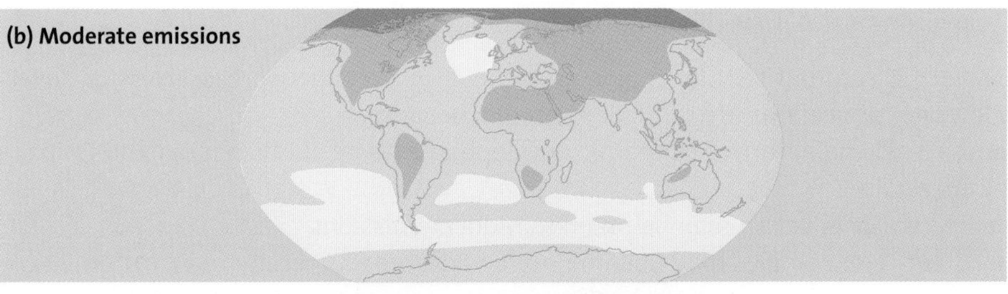

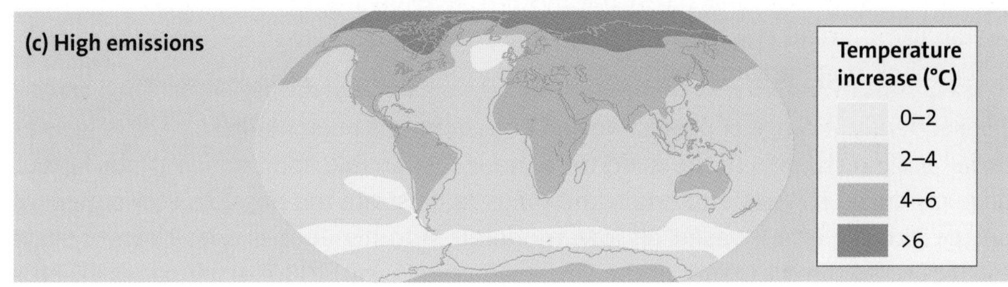

Figure 5.1 *Projected temperature change for the years 2090–99 relative to 1980–99 based on three economic scenarios that assume no new emissions policy*

Table 5.2 Projected greenhouse emissions (present day to 2050)

Emission source	Current emission proportion (%)	Projection (by 2050)
Electricity generation	24	Fastest growing source, especially with rapidly growing population and economic development in India and China
Deforestation	17	Mainly occurring in Indonesia and Brazil; projected to fall
Transport	14	High emissions from road transport and growing emissions from aviation; second-fastest growing source
Industry	14	Major contributions from newly industrialised countries (NICs), but declining elsewhere due to technology and diversified economies
Agriculture	12	Fertiliser use and livestock account for significant emissions; there will be a steady rise as intensive farming is needed to feed the growing world population
Buildings (light/ heat etc.)	8	Will increase by 66% by 2050 because of larger population and urbanisation in developing countries

Source: *Stern Review* (2006)

starting conditions that describe the current state of the climate system accurately, so that they can provide the best possible forecast of the first decade or so, as well as better forecasts further into the future.

More research is therefore needed into:

- the potential risks of large-scale extreme events. It is likely that most impacts will be caused by alteration of the frequency or severity of extreme weather events, where society's capacity to adapt is most limited. Until now, impact studies have concentrated on mean climate conditions, but greater understanding of extreme weather events should also be considered.
- mega-vulnerabilities. Research is needed to understand the cumulative effects of additional stresses (e.g. rising sea levels, AIDS) on climate impacts and adaptation. Climate change will increase the challenges the world faces in providing enough food, energy, clean water and sanitation for everyone in a healthy environment. Dr Robert Watson (the Chair of the IPCC in 2001) stated: 'This is not just an environmental issue, but a development issue.'

Table 5.2 summarises the major projections in greenhouse gas emission levels. However, projections are difficult to predict because:

- the level and nature of economic development, particularly in India and China, is uncertain
- international action may be taken to reduce emissions
- 'green' technologies may become more widespread, thereby reducing emissions
- there is an in-built inertia in the system — even if greenhouse gas emissions stabilise, climate change will continue for several decades
- **feedback mechanism**s, both positive and negative, are not understood fully

Consequently, a range of predictions on emissions has been made from 'business as usual' (i.e. no change) to the need for a more sustainable approach. Various bodies have produced reports, materials and maps to illustrate the possible consequences on the global climate based on low, moderate and high emissions. These include the reports of the IPCC and, in the UK, the *Stern Review* (2006 — see pages 43–46).

In general, scientists agree that, under each of these scenarios, temperature increases on land will be greater than those at sea and that the Arctic will have the most extreme changes. However, the size of the changes is not known and is the subject of speculation.

IPCC Working Group II

In 2007, the IPCC's Working Group II presented its view of the possible impacts of climate change. It identified six sectors of impacts, and eight regions where those impacts would happen (Figures 5.2–5.4).

The six sectors of impact are:

1 **water** — supplies stored in glaciers and snow cover are projected to decline, reducing water availability in regions supplied by meltwater from major mountain ranges, where more than one-sixth of the world population currently lives.
2 **ecosystems** — 20–30% of plant and animal species assessed so far are likely to be at increased risk of extinction if increases in global temperatures exceed 1.5–2.5°C.

***Figure 5.2** Examples of projected global impacts*

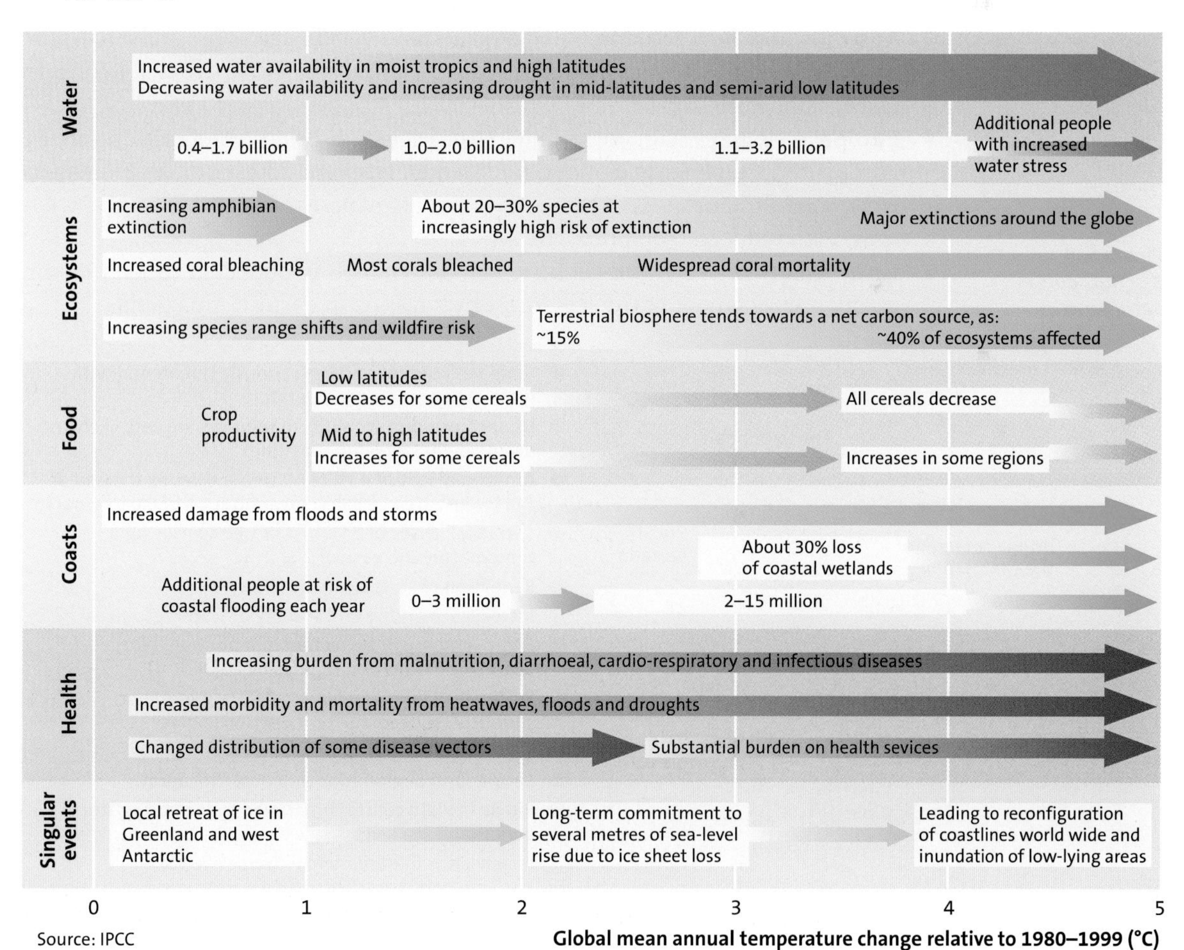

3. **food** — at lower latitudes crop productivity is projected to decrease for even small temperature increases (1–2°C). At higher latitudes crop productivity is projected to increase for temperature increases of 1–3°C, and then decrease beyond that.
4. **coasts** — by the 2080s, many millions more people are projected to be subject to flooding every year due to sea-level rise.
5. **industry, settlement and society** — the most vulnerable industries, settlements and societies are generally those on coastal and river floodplains, those whose economies are closely linked with climate-sensitive resources and those in areas prone to extreme weather events, especially where rapid urbanisation is occurring.
6. **human health** — projected climate change-related exposures are likely to affect the health status of millions of people, particularly those with low adaptive capacity.

The eight areas of impact are:

1. **Africa** — by 2020, 75–250 million people are projected to be exposed to an increase in water stress.
2. **Asia** — projected crop yields could increase up to 20% in east and southeast Asia while they could decrease by up to 30% in central and southern Asia by the mid-twenty-first century.
3. **Australia and New Zealand** — significant biodiversity loss is projected to occur by 2020 in some ecologically rich sites including the Great Barrier Reef and Queensland Wet Tropics.
4. **Europe** — initially, climate change is projected to bring benefits to northern Europe (reduced energy demand for heating, crop and forest growth increases) while southern Europe is expected to experience increased heat waves and wildfires, with reduced crop productivity.
5. **Latin America** — by mid-century, climate change is projected to lead to the gradual replacement of tropical forest by savanna in eastern Amazonia.

Figure 5. 3
Key hot spots in Latin America

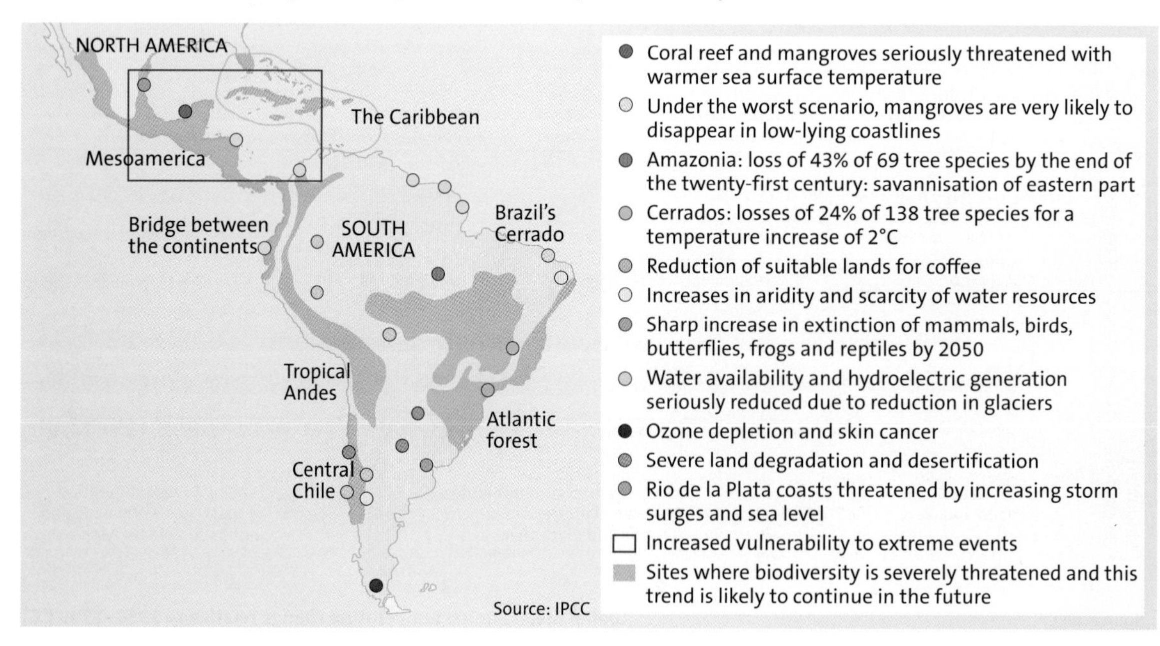

6 **North America** — cities currently experiencing heatwaves will experience many more in the future, with adverse health impacts.
7 **polar regions** — climate change is projected to impact on natural ecosystems with detrimental effects on many organisms including migratory birds, mammals and higher predators.
8 **small islands** — deterioration in coastal conditions, for example through beach erosion and coral bleaching, is expected to affect local resources, e.g. fisheries and tourism.

Figure 5.4 *Projected impacts by region*

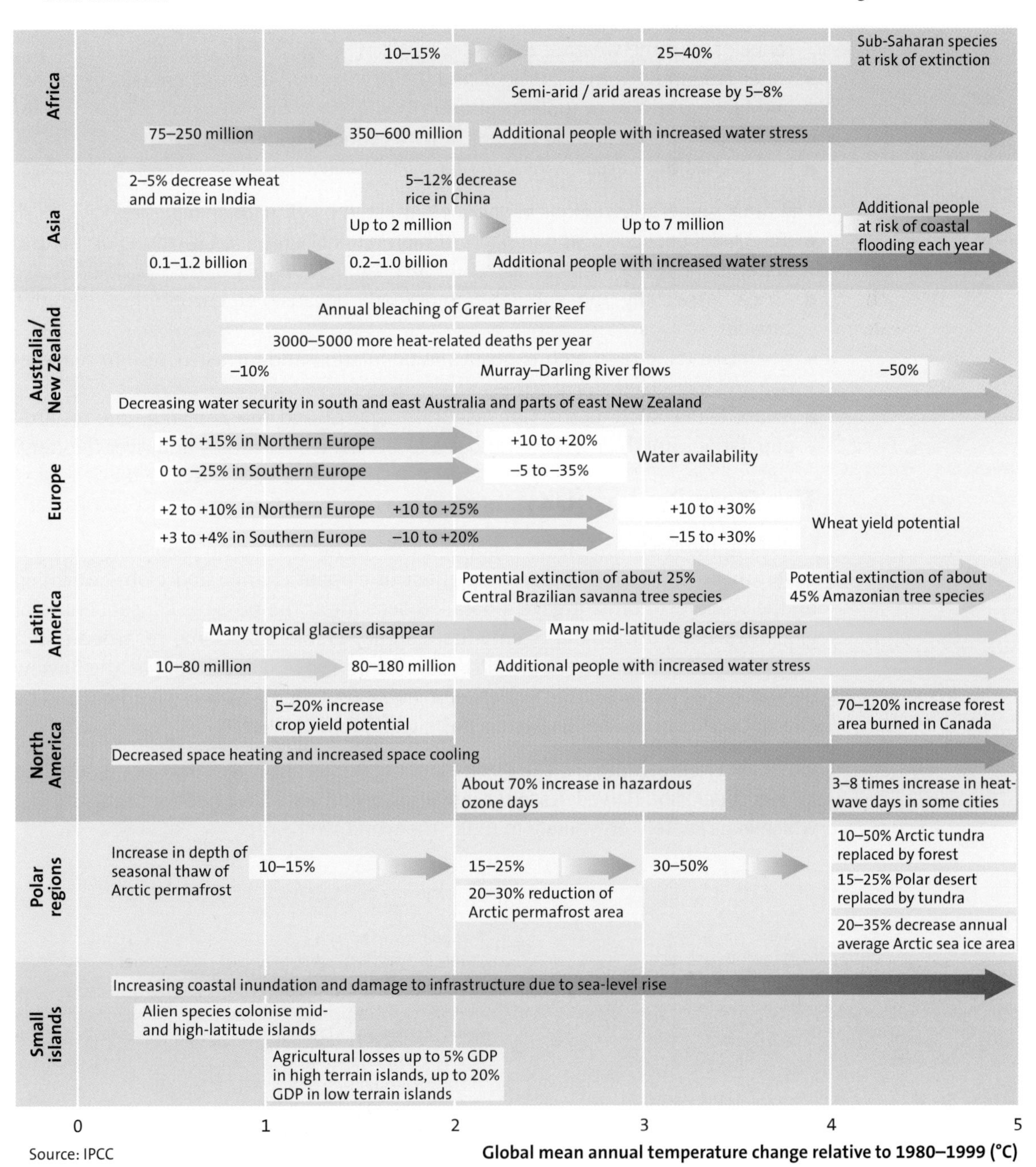

Conclusions

The most vulnerable sectors likely to be affected by climate change are:

- some ecosystems:
  - terrestrial — tundra, boreal forest, mountain, Mediterranean-type ecosystems
  - along coasts — mangroves and salt marshes
  - in oceans — coral reefs and the sea ice biome
- low-lying coastal regions, due to the threat of sea-level rise and increased occurrence of extreme weather events
- water resources in mid-latitudes and the dry tropics, due to decreases in rainfall and higher rates of evapotranspiration
- agriculture in low-latitude regions, due to reduced water availability
- human health in areas with low adaptive capacity

The most vulnerable regions likely to be affected by climate change are:

- the Arctic, because of the impacts of high rates of projected warming on natural systems
- Africa, especially the sub-Saharan region, because of current low adaptive capacity
- small islands, due to high exposure of population and infrastructure to sea-level rise and increased storm surges
- Asian deltas, such as the Ganges–Brahmaputra and the Zhujiang, due to large populations and high exposure to sea-level rise, storm surges and river flooding

The *Stern Review* (UK)

Following the first three IPCC reports, the *Stern Review*, commissioned by the UK government in 2006, discussed the effect of climate change and global warming on the world economy. It gave a detailed account of the possible consequences of global warming during the twenty-first century (Table 5.3). Some regarded it in a very positive manner. The British Prime Minister, Tony Blair, stated that 'the Review demonstrated that scientific evidence of global warming was overwhelming and its consequences would be disastrous if the world failed to act'.

Others have been more critical. Bjørn Lomborg wrote: 'The *Stern Review*...is selective and its conclusion flawed. Its fear-mongering arguments have been sensationalised, which is ultimately only likely to make the world worse off.'

Table 5.3 *An illustration of the main conclusions of the* Stern Review *regarding the effects of temperature increases*

Temperature rise (°C)	Water	Food	Health	Land	Environment	Abrupt and large-scale impacts
1	Small glaciers in the Andes disappear, threatening water supplies for 50 million people	Modest increases in cereal yields in temperate regions	At least 300 000 people each year die from climate-related diseases (diarrhoea, malaria) and malnutrition; reduction in winter mortality in cold countries	Permafrost thawing damages buildings and roads in parts of Canada and Russia	At least 10% of terrestrial species facing extinction; 80% bleaching of coral reefs, including the Great Barrier Reef	Atlantic thermohaline circulation starts to weaken
2	Potential 20–30% decrease in water availability in some vulnerable areas (e.g. southern Africa, Mediterranean)	Sharp declines in crop yields in tropical regions (5–10% in Africa)	40–60 million more people exposed to malaria in Africa	Up to 10 million more people affected by coastal flooding each year	15–40% of species facing extinction; High risk of extinction of Arctic species including polar bear and caribou	Potential for Greenland ice sheet to begin melting, irreversibly, accelerating sea-level rise and leading to an eventual 7 m sea-level rise
3	In southern Europe serious droughts occur once in every 10 years; 1–4 billion more people suffer water shortages; 1–5 billion gain water, which may increase flood risk	15–550 additional millions at risk of hunger; agricultural yields in temperate latitudes likely to peak	1–3 million more people die from malnutrition	1–170 million more people affected by coastal flooding each year	20–50% of species facing extinction, including 25–60% of mammals, 30–40% of birds, and 15–70% of butterflies in South Africa; collapse of Amazon rainforest	Risk of abrupt changes to atmospheric circulations (e.g. the monsoon)
4	Potential 30–50% decrease in water availability in southern Africa and Mediterranean	Agricultural yields decline by 15–35% in Africa; entire regions out of production (e.g. parts of Australia)	Up to 80 million more people exposed to malaria in Africa	7–300 million more people affected by coastal flooding each year	Loss of around half of Arctic tundra	Risk of collapse of the West Antarctic ice sheet
5	Possible disappearance of large glaciers in Himalayas, affecting one quarter of China's population and hundreds of millions in India	Continued increase in ocean acidity, seriously disrupting marine ecosystems and fish stocks		Sea-level rise threatens small islands, low-lying coastal areas (e.g. Florida) and major world cities (e.g. New York, London, Tokyo)		Risk of collapse of Atlantic thermohaline circulation
Over 5	The latest science suggests that the Earth's average temperatures will rise by even more than 5°C or 6°C if emissions continue to grow and positive feedbacks amplify the warming effect of greenhouse gases. The level of global temperature rise would be equivalent to the amount of warming that occurred between the last ice age and today, and is likely to lead to major disruption and large-scale movement of people. Such 'socially contingent effects' would be catastrophic, but are hard to predict with current models as the temperature changes involved are so far outside human experience.					

The views of the World Health Organization

The World Health Organization (**WHO**) states that: 'Climate change is a significant and emerging threat to public health, and changes the way we must look at protecting vulnerable populations.' It has also suggested that the impacts of climate change on human health will not be evenly distributed around the world and has called for research in this area. Figure 5.5 summarises WHO's concerns.

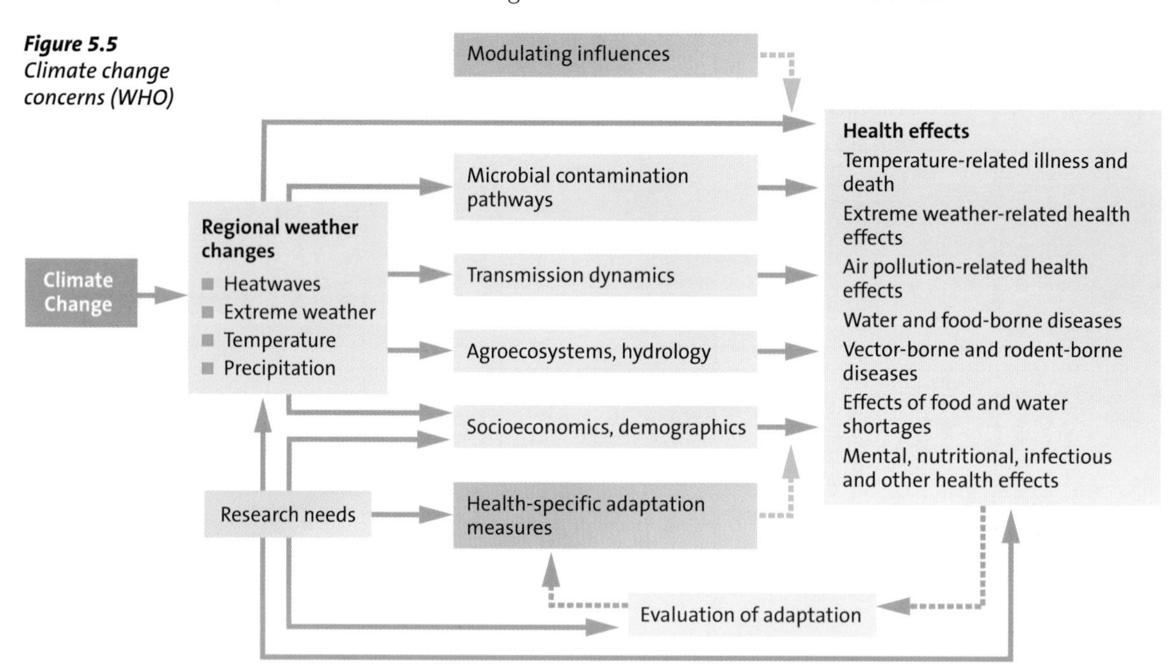

Figure 5.5 *Climate change concerns (WHO)*

Case study 13 THE TIPPING POINT

The **tipping point** is the theoretical point after which the effects of climate change become irreversible. Much of the debate surrounding this is speculative. Passing the tipping point could cause unstoppable changes to the planet. On a global level there could be huge changes in the world's natural systems that could lead to failing infrastructure, mass migration and even war. Impacts at a local scale could include warmer temperatures, flooding or drought. Some like to set a particular °C when this can occur.

Evidence for the tipping point

Evidence began to emerge in 2005 that we may have reached the tipping point. This evidence came from Siberia from the melting of frozen peat bogs that had remained frozen for thousands of years. This would result in the releasing of tonnes of methane (a greenhouse gas) that would further enhance global warming.

Newspapers in the UK featured this worrying sign. In 2008, the *Independent* reported that global warming had reached the point of no return. Areas of Arctic ice had melted in the period 2002–05 to the extent that it was not refreezing, thereby beginning a process of further global warming. As more ocean is exposed to sunlight, the more it warms up and continues the process of further warming. It has been predicted that by 2070 all Arctic sea ice will have disappeared. Other scientists have stated that the levels of carbon dioxide in the atmosphere have reached 400–450 ppm — a critical point beyond which climate change cannot be contained.

Further evidence emerged in 2007:

- Arctic ice was 23% smaller than at any time since records began.
- In Greenland, snow was melting for 25–30 days longer than normal.

Evidence against the tipping point

Conservative estimates state that the Greenland ice sheet could take another 300 years to melt and that changes to global temperatures affecting the Amazon rainforest and Antarctic ice sheet could be at least 100 years away. It is believed by some that there is still time to find solutions to diminish the effects of climate change and that early warnings of tipping points could spur humanity to find solutions and curb emissions. Others believe that the processes involved are still not understood fully, so alarmist claims should be tempered.

Monitoring the situation

There are several areas where further monitoring of the tipping point can take place, including:

- the amount of Arctic sea ice
- the rate of melting of the Greenland ice sheet
- the rate of melting of the West Antarctic ice sheet
- the rate of changes to the El Niño effect
- the timings of the Indian and West African monsoons
- the rates of tree death in the boreal forests of Siberia and Canada

The UN Secretary General, Ban Ki-moon, said in 2007 that the effects of climate change have become 'so severe and so sweeping that only urgent global action will do. There is no time to waste.'

11 Using case studies

Question

In February 2008, the journal *Proceedings of the National Academy of Sciences* suggested that a variety of tipping elements could reach their critical points within this century, under anthropogenic (man-made) climate change:

- **Arctic sea ice — some scientists believe that the tipping point for the total loss of summer sea ice is imminent.**
- **Greenland ice sheet — total melting could take 300 years or more but the tipping point that could see irreversible change might occur within 50 years.**
- **West Antarctic ice sheet — scientists believe that it could collapse unexpectedly if it slips into the sea at its warming edges.**
- **Gulf Stream — few scientists believe it could be switched off completely this century, but its collapse is a possibility.**
- **El Niño — the southern Pacific current may be affected by warmer seas, resulting in far-reaching climate change.**

- **Indian monsoon — relies on the temperature difference between land and sea, which could be tipped off-balance by pollutants that cause localised cooling.**
- **West African monsoon — in the past it has changed, causing the greening of the Sahara, but in the future it could cause droughts.**
- **Amazon rainforest — a warmer world and further deforestation may cause the collapse of the rain supporting this ecosystem.**
- **Boreal forests — cold-adapted trees of Siberia and Canada are dying as temperatures rise.**

Source: The *Independent*, 5 February 2008

How helpful is such reporting in the climate change debate?

Guidance

This question asks you to reflect on whether speculating on possible doomsday scenarios (as given in the bullet points above) encourages people to change their habits, or rather to sit back and think that there is little point in doing anything. Indeed, is the alarmist language being used contributing to making the situation self-fulfilling?

Coping with climate change

This section of the book examines the various ways in which international bodies, national governments, local governments, industries and individuals are trying to cope with the onset of climate change. In most cases, people are persuaded that something can be done to prevent the worst effects — there is still time to do something.

There is global agreement that climate change needs to be addressed, but much less agreement on how this can be achieved. In general terms, there are two approaches: **mitigation** and **adaptation**.

Mitigation refers to a reduction in the output of greenhouse gases and/or increasing the size and amount of greenhouse gas storage or sink sites. Examples of mitigation include:

- setting targets to reduce greenhouse gas emissions
- switching to renewable sources of energy
- 'capturing' carbon emissions and/or storing or burying them

Adaptation refers to changing our lifestyles to cope with a new environment rather than trying to stop climate change. Examples of adaptation include:

- developing drought-resistant crops
- managing coastline retreat in areas vulnerable to sea-level rise
- investing in better-quality fresh water provision to cope with higher levels of drought

It is argued by some that no matter what we do in the twenty-first century to mitigate climate change, some aspects of adaptation will still be required. Mitigation and adaptation are not alternatives — they will have to operate together.

There is also a key difference between the ability of human systems and the ability of natural systems to respond to both mitigation and adaptation.

For human systems, mitigation involves costs upfront to, say, reduce atmospheric pollution levels. Adaptation involves costs over a longer timescale as the effects of climate change emerge. There is also the issue that some areas of the developed world may have the resources to adapt to changing climate, whereas others in the poorer developing world lack the adaptive capacity (i.e. resources) to cope. Ability to adapt is therefore linked to development. Most adaptation strategies will be local in scale, as adaptations need to be tailored to local impacts of climate change. On the other hand, mitigation can operate at a variety of scales:

- individual — lifestyles and consumption choices
- local — local government strategies on planning, recycling and transport
- national — government policies and national tax frameworks
- global — international agreements for global action

For natural systems, mitigation will limit damage to ecosystems. On the other hand, adaptation may condemn some elements of natural systems that cannot adapt, and they will disappear or become extinct. The value of natural systems forms a strong argument for acting now to reduce the worst impacts of climate change. Much work is now being done on the creation of large reserve-aligned natural systems with wildlife corridors to permit species movement.

Table 6.1 illustrates a range of climate coping strategies, across a spectrum from adaptation to mitigation. All such strategies have costs and benefits.

Adaptation involves a range of strategies. The United Nations Environment Programme (UNEP) recognises eight possible adaptation options. Table 6.2 evaluates these strategies.

The case studies in this section examine a range of mitigation and adaptation strategies around the world and at a variety of scales.

Table 6.1 Climate coping strategies

	Scheme	Examples of nature of scheme	Advantages	Disadvantages
Mitigation	Carbon-neutral development	Offsetting of all carbon emissions through afforestation; wholesale switching to renewable energy	Has a fundamental impact on emissions	Could prevent development; public opposition to land use and lifestyle changes
	Carbon-capture technology	Applied to power stations and industry with deep sea or geological burial	Removes the problem at source	High costs passed on to consumers; encourages continued use of finite resources
	Sustainable development	Reduced resource consumption; increased recycling of materials; locally sourced foodstuffs; alternative transport	Some aspects have already been accepted by the public	Lifestyle changes may be opposed, and the changes are slow to take effect
	Geoengineering	Orbiting solar shields to reflect incoming solar radiation; ocean iron seeding to increase algal growth and sequester carbon	Could provide a 'one-off' solution and avoid lifestyle changes	Huge costs and untried technology; side-effects largely unknown
	Agricultural technology	Drought-tolerant crops; no-tillage systems; water-harvesting technology; the use of waste water on fields	Much of the technology already exists and could be adopted easily	Costs may be prohibitive in the developing world, where the need is greatest
Adaptation	Land-use planning	Preventing development on floodplains and vulnerable coasts; removal of urban scrubland to prevent the spread of fire	Reduces vulnerability to extreme weather events	Costly to implement and may be opposed by existing residents and businesses; small scale only

Table 6.2 Adaptation strategies

Strategy	Description	Appraisal
Bear the loss	As costs occur (e.g. due to increased flooding), they are absorbed	The costs may become unbearable if they occur too frequently
Share the loss	As costs occur, relief is provided by governments, aid agencies and insurers	The costs may become too large for society and the economy to cope with
Modify the threat	The costs are reduced by some form of protection (e.g. a flood barrier)	If the threat grows further, costly investment may be needed
Prevent	One response to increased drought could be to develop a new water source, such as drip-fed irrigation	The investment costs are likely to be high
Relocate	Abandon areas for less risky locations	This assumes that there are new areas available
Research	Investigating the problem and developing new technological solutions (e.g. drought-tolerant crops)	This requires technical expertise, funding and infrastructure
Change use	Different crops grown in response to changing climate or developing hill walking holidays in former ski resorts	This involves long-term planning and reinvestment costs
Change behaviour	Education, for example to encourage people to conserve water	A long-term option; producing lifestyle change is notoriously difficult

THE KYOTO PROTOCOL

Case study 14

Attempts globally to mitigate climate change have centred on the 1997 Kyoto Protocol. The origins of the protocol were in 1992 at the Earth Summit held in Rio de Janeiro. Here, the United Nations Framework Convention on Climate Change (UNFCCC) was endorsed by 186 countries. Its aim was to achieve stabilisation of greenhouse gases in the atmosphere in order to prevent dangerous anthropogenic interference with the climate system.

In 1997, at a follow-up meeting in Kyoto, Japan, over 100 governments signed a 'Climate Change Protocol'. This set more specific legally binding targets for pollution mitigation and proposed schemes to enable governments to reach these targets. Most governments agreed that by 2010 they should have reduced their atmospheric pollution levels (of greenhouse gas emissions) to those of about 1990. The Kyoto Protocol came into force in February 2005 and by 2006 had been ratified by 183 countries. However, this agreement will expire in 2012.

The goal of the Kyoto Protocol is to reduce worldwide greenhouse gas emissions between 2008 and 2012 to 5.2% below 1990 levels. Compared with the emissions levels that would occur by 2010 without the Kyoto Protocol, this target actually represents a 29% cut.

The Kyoto Protocol sets specific targets for emissions reduction for each industrialised nation, but excludes developing countries. To meet their targets, most ratifying nations would have to combine several strategies:

- place restrictions on their biggest polluters
- manage transportation to slow or reduce emissions from vehicles
- make better use of renewable energy sources — such as solar power, wind power, and biodiesel — and using them in place of fossil fuels

Most of the world's industrialised nations support the Kyoto Protocol. One notable exception is the USA, which releases more greenhouse gases than any other nation and

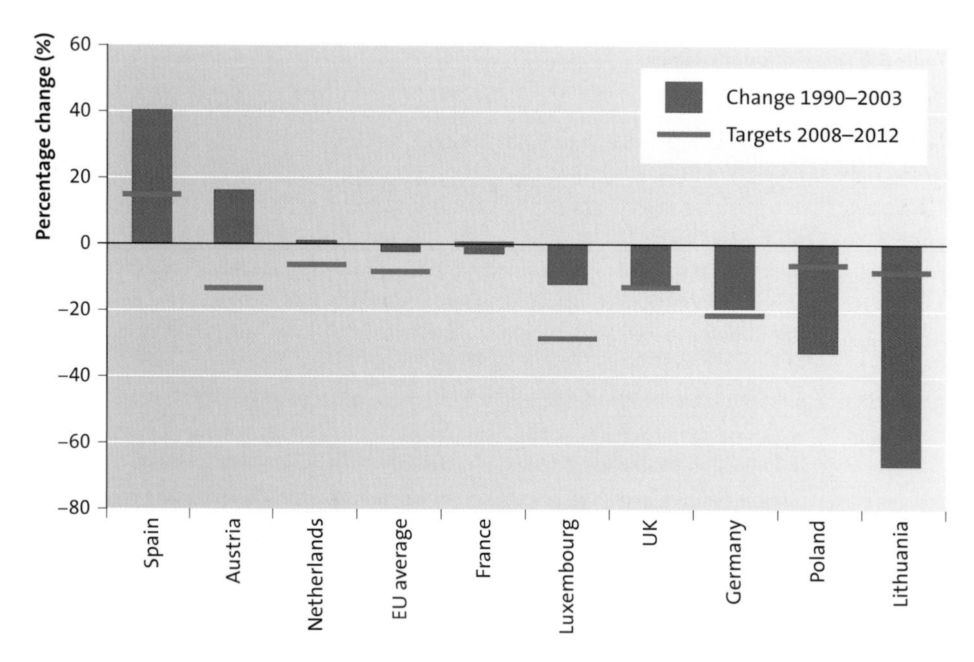

Figure 6.1 *Percentage change in greenhouse gas emissions (selected countries), 1990–2003*

accounts for more than 25% of those generated by humans worldwide. Australia also declined to ratify it until 2007.

Progress on the protocol has been complex for a number of reasons:

- Under terms of the agreement, the Kyoto Protocol would not take effect until 90 days after it was ratified by at least 55 countries involved in the UNFCCC. Another condition was that ratifying countries had to represent at least 55% of the world's total carbon dioxide emissions for 1990. The first condition was met on 23 May 2002, when Iceland became the 55th country to ratify the protocol. When Russia ratified the agreement in November 2004, the second condition was satisfied and the Kyoto Protocol entered into force on 16 February 2005.
- Emissions reduction targets are country specific. Some key goals for individual states and groups include 8% for the EU, 7% for the USA, 6% for Japan, 0% for Russia, and allowed increases of 8% for Australia and 10% for Iceland. Figure 6.1 shows the percentage change in greenhouse gas emissions (1990–2003) compared with the agreed Kyoto targets for 2008–2012 for selected countries.
- Complex systems were introduced to allow the 'trading' of carbon credits, i.e. buying unused emissions from other countries. Carbon sinks, such as planting forests are allowed, so that countries can 'offset' emissions. Critics state that both of these allow polluters to continue to pollute.

The Kyoto Protocol has not been a great success in reducing emissions of greenhouse gases. Between 1992 and 2007, global emissions in greenhouse gases increased by 38% (see Figure 6.2). China's emissions increased by 150%, India's by 103%, the USA's by 20% and Japan's by 11%. Within the EU, there has been an overall fall of 0.8% from the group known as EU-15 (the members prior to 2004) with significant falls from the UK, Germany, Denmark and Sweden, but this is still below the target.

In 2005, the Gleneagles Action Plan was devised, which included plans for improved energy efficiency in buildings and appliances, cleaner fuels, renewable energy and the

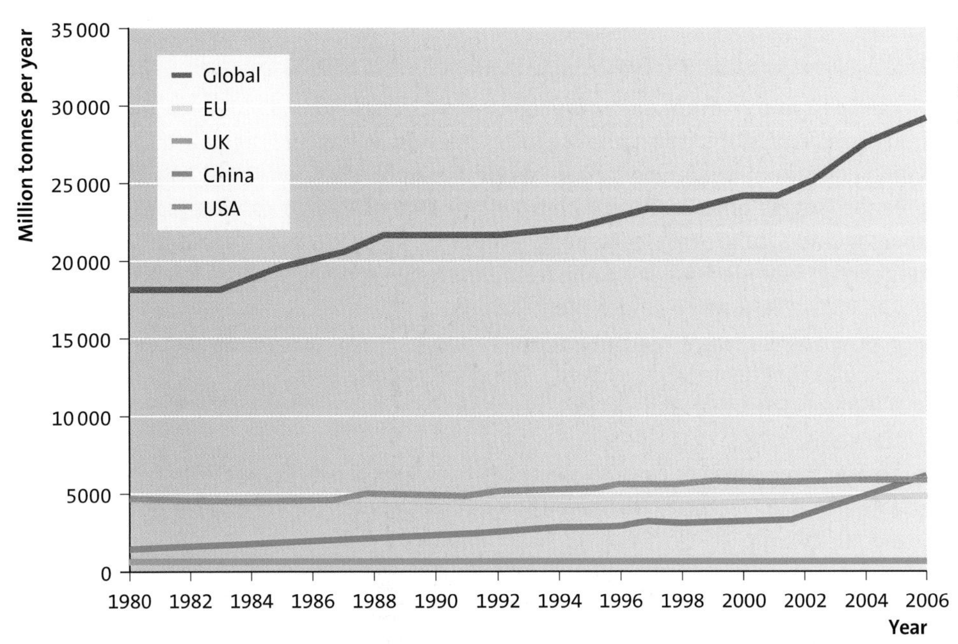

Figure 6.2 *Changes in global carbon emissions, 1980–2006*

promotion of research and development into new cleaner technologies. More cuts were agreed at the UN Climate Change convention in Bali in 2007 (the so-called 'Bali road map'), with the final details to be decided in Copenhagen in December 2009. It is hoped that this agreement will succeed the Kyoto Protocol. It is also hoped that the US government, under the administration of President Barack Obama, has now recognised that climate change is a threat.

12 Using case studies

Question

Draw up a brief five-point plan setting out your own ideas for a post-2012 'son of Kyoto' international agreement on climate action.

Guidance

Remember the three main tenets by which the original Kyoto Protocol sought to set targets to reduce emissions:

- to place restrictions on their biggest polluters
- to manage transportation to slow or reduce emissions from vehicles
- to make better use of renewable energy sources — such as solar power, wind power and biodiesel — in place of fossil fuels

Now, you could factor in two further considerations:

- the rise of emissions from India and China, both of which were exempt from the Kyoto Protocol — how should we deal with these two emergent superpowers?
- an evaluation of the success of schemes such as **carbon trading** and **carbon offsetting**, both of which have arisen and have been criticised for allowing industrial nations to continue to pollute

Remember to make your five-point plan SMART (specific, measurable, attainable, realistic and with time limits), and easy to understand, for media purposes.

Case study 15 MITIGATION AND THE EU

Under the Kyoto Protocol, the European Union (EU) agreed to an 8% reduction in greenhouse gas emissions compared with 1990 levels. It has also agreed to a '20/20/20 vision', a form of energy policy for Europe, whereby there will be a 20% reduction in greenhouse gas emission, a commitment to 20% of energy coming from renewable sources, and a 20% increase in energy efficiency, all by 2020. As part of this strategy, a target of 21% of electricity to come from renewable sources by 2010 has also been set.

The 20/20/20 vision includes the following:

- a renewable energy road map, which describes various renewable energy technologies that could be developed more widely across Europe
- a Biofuels Directive that establishes targets for the proportion of biofuels used in transport
- greater integration of gas and electricity networks across Europe
- a Renewable Electricity Directive that sets targets for the renewable portion of the electricity supply
- a carbon-trading system called **EUETS** (European Union Emission Trading Scheme — Figure 6.3), which is a key market mechanism for driving down industrial carbon emissions through a cap and trade system

EUETS is the world's only compulsory example of a 'cap and trade' system — a mechanism that sets limits (cap) on the emission of a pollutant but allows companies that are within the limit to sell credits (trade) to companies that need to pollute more. This covers around 50% of all EU carbon emissions. The power generation, steel, cement and other heavily polluting industries are part of the scheme. It operates as follows:

- The EU, each nation within it, and each carbon producing installation within each country is given a cap.
- In 2005, the UK had a cap of 736 million tonnes (mt) of carbon dioxide.
- Each of the 1078 installations (power generation, steel, cement etc.) within the UK had its own cap.
- Consider installation A: it has a cap of 1 mt, and produces 1 mt of emissions, so no further action is needed,
- Consider installation B: it has a cap of 0.5 mt yet produces 0.6 mt of emissions. This installation must either increase energy efficiency or buy a carbon credit to the value of 0.1 mt.

Figure 6.3
EUETS

EU target emissions set at 6572 million tonnes (mt) of carbon dioxide for 2005–07 → UK 'share' set at 736 million tonnes — the UK's 'cap' → 1078 UK installations, such as power stations and cement works, are included in the scheme. Each is given its own 'cap'

Emissions = 1 mt
No action needed

Emissions = 0.6 mt
Must either increase efficiency, or buy carbon credits

Emissions = 0.6 mt
Sells 0.2 mt to the EU carbon market = profit

- The EU cap, national caps and installation caps are set to achieve the Kyoto targets
- Phase 1 ran from 2005 to 2007
- Phase 2 will run from 2008 to 2012

Figure 6.4
A cynical view of carbon trading?

- Consider installation C: it has a cap of 0.8 mt and produces 0.6 mt of emissions. This company can sell 0.2 mt of carbon credit to the EU, and hence profit from its energy efficiency.
- During 2009, the ETS credits were valued at £11 per tonne — some experts believe they are severely undervalued and should be traded at nearer £100 per tonne if they are to be more effective in cutting emissions from the big polluters.

Progress towards the 20/20/20 vision is variable. In 2006, the EU was at 7.7% greenhouse gas emissions reduction (including only a 0.2% reduction in carbon dioxide), 9.2% of energy from renewable sources (including 14.6% renewable electricity) and a 17.6% reduction in energy intensity (i.e. greater efficiency).

MITIGATION AND THE UK

Case study 16

The *Stern Review* (2006) (see p. 46) concluded that if climate change is allowed to continue unchecked it would cost at least 5% and possibly 20% of global annual GDP. If the world decided to mitigate climate change, the cost would be about 1% of GDP per annum. Central to Stern's argument is that it is more sensible to invest in mitigation today than to do nothing and face spiralling costs and global recession. Stern also concluded that emissions should be stabilised at between 450 ppm and 500 ppm carbon dioxide, and that this would require global emissions to peak in the next 10–20 years, and then fall at a rate of at least 1–3% per year. In response to this review, the UK government introduced the Climate Change Act of 2008.

The act embedded a number of aspirational targets within UK law. They include:

- an 80% reduction of greenhouse gas emissions by 2050 (compared with 1990 levels)

- a minimum of a 26% reduction of carbon dioxide emissions by 2020 (including a 20% reduction of carbon dioxide by 2010)
- a commitment to 10% renewable electricity by 2010 and 20% by 2020

How can these targets be achieved?

The UK government is looking towards the following methods to achieve these targets:

- a **renewable obligation certificate** (ROC) regime whereby energy suppliers face an increasing requirement year on year to supply a rising proportion of renewable energy, initially targeting 20% by 2020. This is expected to create a massive expansion of wind power, both onshore and offshore.
- possible support for a tidal barrage across the River Severn at a cost of around £30 billion to produce about 4% of the UK's electricity
- possible feed-in tariffs for micro-renewable electricity from, for example, small-scale solar panel users. Village grants are now available for communities to sell surplus electricity from wind farms.
- a new renewable heat strategy is under development, which may involve more use of recycled heat in, for example, new housing and office developments
- a reliance on EUETS (see above) to reduce carbon emissions through increased investment and/or allowance of trading
- massive private investment in new generation capacity: nuclear (to replace at least the 11 GW — up to 20% of supply — we have currently from nuclear power stations that are to be decommissioned) and new gas and coal-fired capacity that is to be carbon capture ready
- further interconnection with the EU (and wider) markets. For example, the UK now has three LNG (liquid natural gas) terminals (Isle of Grain, Teesport, Milford Haven), two major gas interconnections to Norway and Belgium and a second electricity cable link to the Netherlands coming online in addition to the existing 2000 MW UK–France line.

Progress so far?

The UK is currently (2008 data) at an 8.5% carbon dioxide reduction, a 22% greenhouse gas reduction, and a 1.5% share from renewable energy sources, including a 4.6% share from renewable electricity. There is clearly some way to go in meeting the targets. Indeed they are already stimulating debate:

> To begin to meet the legal targets of the Climate Change Act, the UK will have to achieve and maintain decarbonisation at (unprecedented) rates. The Climate Change Act will have to be revisited by Parliament or simply ignored by policymakers. What are the costs, in terms of public cynicism about legislators and the legislative process, of passing aspirational, rather than codifying, laws?
>
> Professor Gwyn Prins (London School of Economics)

> The task (of cutting emissions by 80% from 1990 levels by 2050) is already staggeringly huge and, as we have seen, well beyond our current political capacity to deliver. It is hard to see any tough choices being made in the current climate.
>
> Colin Challen MP, chairman of the All Party Parliamentary Climate Change Group

The UK's progress towards meeting these targets by 2050 will create a wide range of issues. If we are to meet our goals, at least £100 billion investment in the UK energy sector will be needed over the next 20 years. Tens of thousands of jobs could be created to deliver the infrastructure development needed, especially in the key sectors of

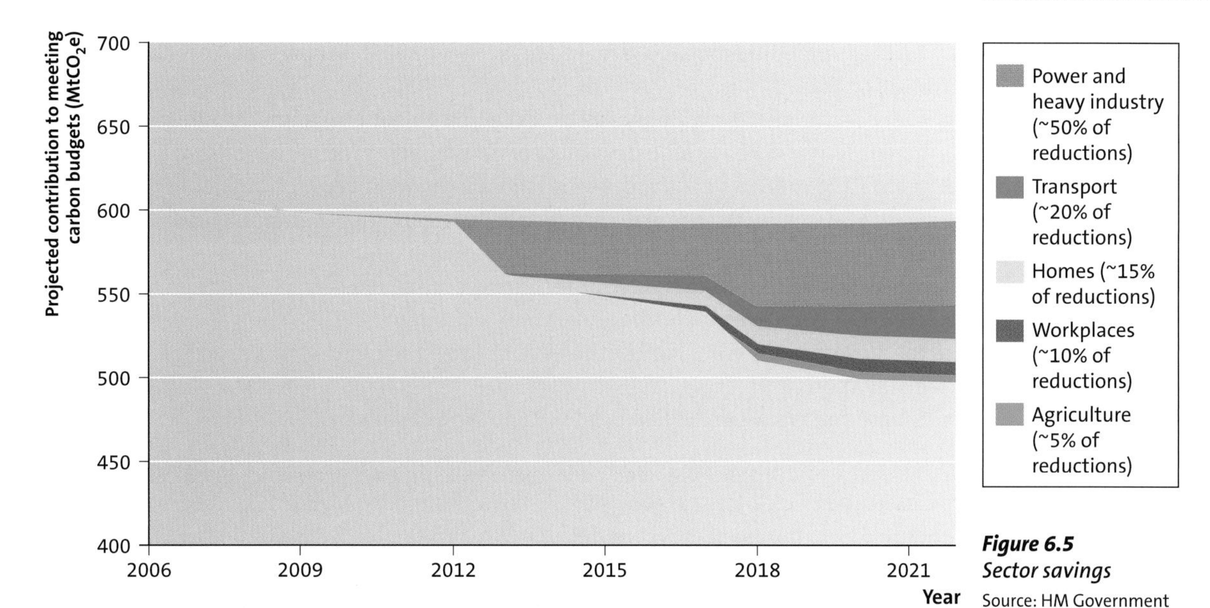

Figure 6.5
Sector savings
Source: HM Government

nuclear power decommissioning, waste management and development, renewable sources, gas and electricity transmission, energy engineering and research and development. A key political aim will be to increase energy security, so that the UK relies less on imported sources.

In April 2009, in the budget, the UK government strengthened its resolve in this respect. It also announced:

- a legally binding carbon budget committing the UK to cutting carbon dioxide emissions by 34% by 2020
- incentives for offshore wind farms (subsidies and capital allowances)
- £750 million investment to research into low-carbon technology
- funding for four trial carbon capture schemes (CCS) on coal-fired power stations (e.g. Kingsnorth, Hatfield, Ferrybridge — see below)
- up to £4 billion in loans for renewable energy projects from the European Investment Bank
- a review of the North Sea tax regime to stimulate investment in new fields
- new combined heat and power plants to be exempt from the climate change levy from 2013

In the view of some, the government has hitched the country's future to wind energy, even though the technology is still in its infancy. Only one company (Siemens) makes offshore turbines. The turbines have to be resilient to perform in the stormy conditions at sea. Vestas, the world's largest turbine maker, has pulled out of the European market after a disastrous project at Horns Rev on the Danish coast where it had to bring all 80 of the turbines it had installed back to shore to fit new gearboxes. In addition, maintenance boats have to be available at short notice to go out and fix any problems, and there are few such boats available. The cost of offshore farms is roughly twice that of onshore farms. Is the added expense offset by the additional power produced? Evidence suggests that existing offshore farms operate at full power for only 29% of the time. Twenty per cent of the time they are in need of maintenance (compared with 5% for onshore wind farms).

Further announcements were made in July 2009, as part of the Low Carbon Transition Plan:

- every sector of the economy will be expected to cut emissions, although electricity generation and heavy industry will bear about half of the reductions (see Figure 6.5)
- up to £120 million to advance the offshore wind industry
- up to £60 million to stimulate progress in wave and tidal technologies
- £6 million to explore geothermal energy potential
- a new facility to research nuclear technology
- financial incentives for home generation
- the government will also use powers to speed up grid connection for renewable installations
- 'smart' meters are to be deployed in 26 million homes by 2020

While the UK's renewables target has now risen to a 30% share of the electricity sector by 2020, the low-carbon target is 40% — the difference implying a 10% share for nuclear energy. Seven thousand new wind turbines may have to rise from the land and the sea by 2020. According to Climate and Energy Secretary Ed Miliband, resistance to the technology would have to change. He said: 'It is important to be sensitive to people's issues around wind power, but our default position as a country needs to change. The biggest threat to our beautiful countryside isn't wind turbines, it's climate change.' Ministers also say that the plan puts the UK in a leadership role in the months leading up to the important UN climate summit in Copenhagen in December 2009.

13 *Using case studies*

The UK government is piloting a system in which every government department and ministry will have its own carbon budget. It is proposed that by 2020 there will be a 34% reduction in average annual emissions required by these budgets. The plan will deliver emissions cuts in all sectors through a varied set of policies, ranging from the EUETS capping emissions from heavy industry, supporting innovative new low-carbon technologies, as well as helping people make low-carbon choices in their homes, their workplaces and when they travel:

- **For the Department for Children, Schools and Families (DCSF) it is equivalent to around 1 million families getting their children to school by walking instead of taking them in the car.**
- **For the Ministry for Communities and Local Government (CLG) it is equivalent to eliminating all emissions from 1 million homes.**
- **For the Department for Transport (DfT) it is equivalent to around 4 million people cycling five miles to work instead of taking the car.**

Choose one of these and suggest how it could be made to work.

Clean coal

Carbon capture and storage (CCS — Figure 6.6) has been used so far only in small power stations that generate about 30 MW of electricity, but the new coal power plants would produce 50 times as much power. However, plants fitted with CCS equipment require more coal to operate, using up to 25% to 40% of the power generated to run the carbon-capture equipment, meaning that more fuel has to be

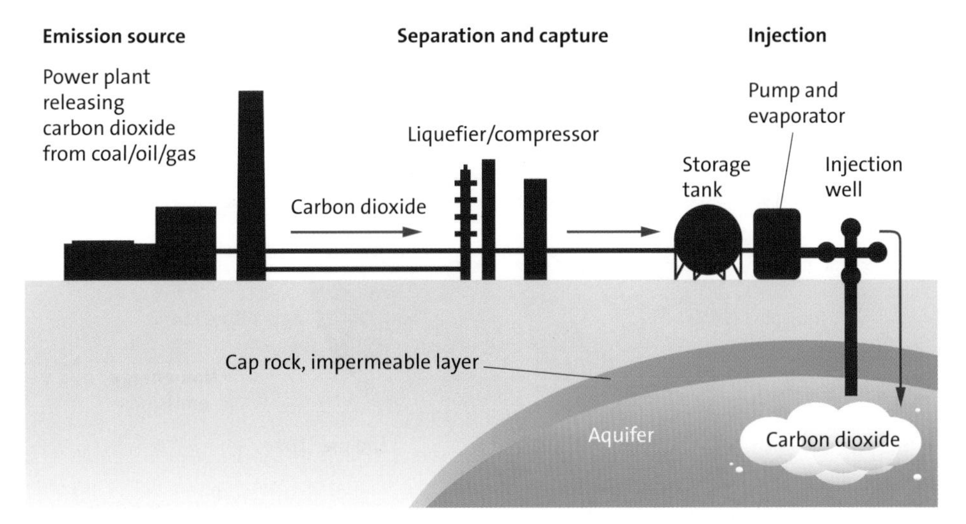

Figure 6.6 *'Clean coal': how carbon capture works*

burnt to produce the same amount of electricity. In addition, it is thought that they should operate with pre-combustion technology in which 90% of carbon emissions are captured.

The government wants to experiment with different types of CCS. Of the four new coal-powered stations planned, at least one would be fitted with the pre-combustion system. Another would be fitted with a post-combustion system in which the carbon is removed after the coal is burnt, capturing only 20% to 25% of emissions. In each case, the carbon dioxide is cooled until it is a liquid and then piped or shipped away for injection into underground stores, typically old oil or gas fields.

Under the government's proposal, no new coal-fired power station will be licensed unless at least 400 MW of gross capacity is fitted with CCS. However, this represents only about 25% of the total generating capacity of proposed coal plants. The government is relying on CCS working by 2025, but it could take longer. Critics also state that more research should be carried out into ensuring that the stored carbon dioxide will not leak out.

The role of businesses in managing climate change

For many years the large oil companies and car manufacturers seemed to have a vested interest in polluting: cleaning up pollution costs money, while polluting costs nothing. Figure 6.7 illustrates the main sources of greenhouse gases in 2000, by economic sector. It is clear that a large number of companies in a wide range of economic sectors potentially play an important role in the emission of greenhouse gases.

Today, the great majority of companies have changed their attitude to climate change. Indeed, many now see it is an economic 'opportunity'. The changing attitude of businesses can be explained by a range of factors, including:

- moral and public pressure to protect, and not destroy, the environment

Figure 6.7
Source of greenhouse gas emissions in 2000 by economic sector
Source: *Stern Review*

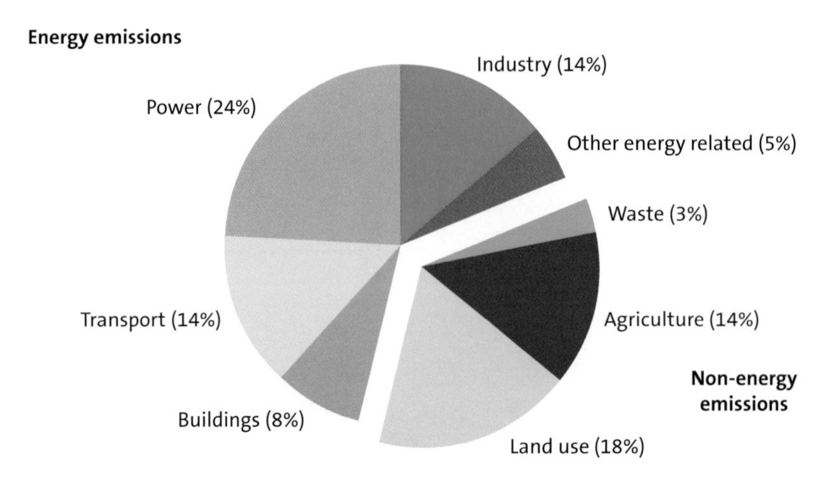

Total emissions in 2000: 42 $GtCO_2e$
Energy emissions are mostly CO_2 (some non-CO_2 in industry and other energy related). Non-energy emissions are CO_2 (land use) and non-CO_2 (agriculture and waste).

- fears about energy supply, linked to high energy prices
- increased moves by governments towards taxing carbon emissions
- demands from investors, such as pension funds, that companies should be environmentally sound
- new technologies, such as renewable energy, hybrid cars and energy-efficient appliances that represent new markets

In 2007, the US Climate Action Partnership was established. This group of businesses, which included BP America, Shell, General Motors and Ford, called for 'a pathway that will slow, stop and reverse the growth of US emissions while expanding the US economy'. It is clear that the development of renewable energy technologies, hybrid cars and energy-efficient buildings, each of which requires research and development, will provide new jobs and new opportunities for economic growth. Are businesses changing their attitudes because, rather than seeing their profits decline, they might actually increase?

Local responses to climate change

Sustainable development has become a key concept in the twenty-first century. In order to maintain current standards of living and to slow down the predicted impacts of global warming, societies need to use their resources more wisely. At the regional scale, many local governments have developed initiatives to respond to environmental problems such as global warming. Some cities are planning to follow London's lead and use a congestion charge to limit the traffic in their centres, although the people of Manchester voted to reject such a scheme. Most of the initiatives introduced in our cities to help reduce problems of urban pollution, for example tram services and park-and-ride schemes, will also help to curb the rise in carbon emissions. In recent years, many UK communities have been faced with increased risk of flooding and this is likely to continue, so it is acknowledged that people should begin to 'think globally and act locally'. It is widely accepted that much energy is wasted in the home. Increased understanding is needed in relation to energy conservation and recycling. If people become more aware of the ways in which the problems can be tackled, then everyone can play their part. Education has an important role to play in this, but individuals can respond to global warming at the local level by:

- improving home insulation. Many local councils assist with the cost of cavity wall insulation and loft lagging. New housing is governed by regulations that ensure properties are properly insulated, so it is older housing that is covered by grants to assist with the cost of these types of home improvement. Double glazing also helps to reduce heat loss and is a feature of most new housing.
- recycling. Individuals can manage their refuse so that recyclable items are separated from non-renewable items. Most local councils have targets for recycling and collect garden waste separately from household waste. Householders are also encouraged to sort out paper and cardboard, glass and aluminium cans, and many councils provide recycling boxes so that this waste can be collected from people's homes.
- using energy wisely. Individual households can cut energy bills significantly by adopting a few simple practices, such as turning off lights, keeping the central heating thermostat at a lower temperature, turning the television and computer off properly, and by using energy-efficient light bulbs (which became EU law in September 2009).
- using public transport and joining car-sharing schemes or, better still, walking to work or school. A number of primary schools have promoted 'walk-to-school' schemes which have been encouraged and advertised in local newspapers and on television. In the Low Carbon Transition Plan, the DCSF (the Department for Children, Schools and Families) has been set a target equivalent to around 1 million families walking their children to school instead of taking them in the car.
- calculating our individual carbon footprint (there are several such calculators on the internet). This may enable us to see which behaviours have most impact and what we can change.

Some of these actions seem simple and small in scale but when added together for millions of people and households they could have a significant effect. Even here though there are differences of opinion. It is said if every light bulb in London was energy efficient, then 575 000 tonnes of carbon dioxide emissions per year would be saved. On the other hand, people are encouraged to switch off their mobile phone chargers, yet all the energy saved in switching off a charger for 1 day is used up in 1 second of car-driving. The energy saved in switching off the charger for 1 year is equal to the energy needed for a single hot bath. Some ask, is it worth it?

There are also problems with the carbon footprint approach. Asking everyone to reduce their footprints could be seen as unfair because some people consume far less than the average. Millions of people in the developing world do not consume enough to have even a reasonable quality of life. There is a strong case for them to consume more.

Case study 17 LONDON AND CLIMATE CHANGE

London is responsible for 8% of the UK's total greenhouse gas emissions. Given London's forecast economic and population growth, this will increase to 15% by 2025.

In 2007, the Mayor of London launched the 'Action Today to Protect Tomorrow' plan for London. The core message of the plan is that:

> Londoners do not have to reduce their standard of living for London to play its part in tackling climate change, but we do all have to change the way we live. We have to move from a high energy-use, wasteful economic model to one that conserves energy and minimises waste. In other words, we have to be more efficient. As our focus is on efficiency, many of the measures advocated in this plan will deliver net financial benefits over a relatively short period of time, as well as cutting emissions.

The plan commits the city to reducing its carbon dioxide emissions to 30% of 1990 levels by 2025. Various initiatives were proposed, including:

- the Green Homes Programme — subsidised, or free, home insulation; improving energy efficiency in existing homes; a website for information and a helpline for advice on energy efficiency in the home

Figure 6.8
The Greater London Low Emission Zone

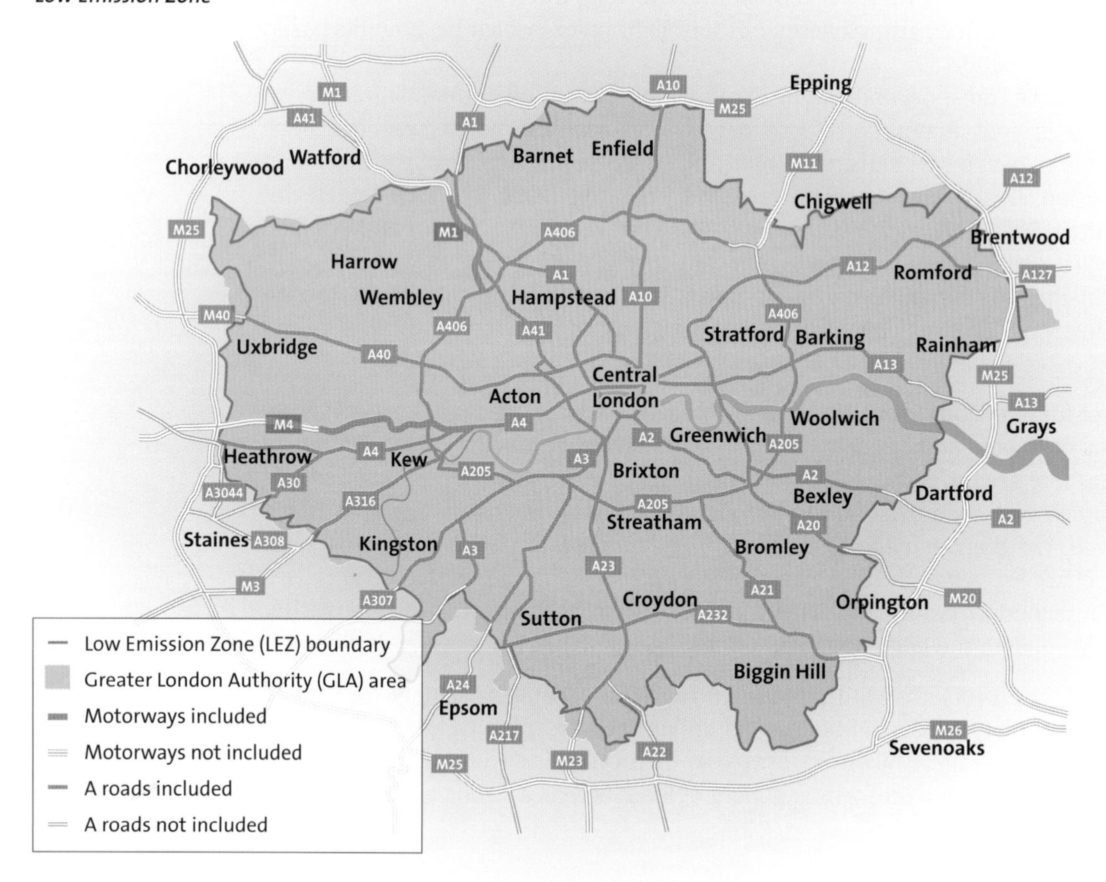

- setting and enforcing new building standards for energy efficiency
- investing in local, small-scale, renewable energy schemes such as solar and wind power
- encouraging 'waste-to-energy' schemes as an alternative to landfill
- providing clean, efficient public transport (converting all 8000 London buses to diesel–electric hybrids); raising the London congestion charge for heavy polluting vehicles; encouraging cycling
- promoting the purchase of low-carbon goods and services by council offices and other bodies

The Greater London Low Emission Zone (LEZ — Figure 6.8) was introduced in February 2008. This is a specified area in Greater London within which the most polluting vehicles are required to pay a daily charge. Although initially only applying to lorries, by 2012 it will apply to all heavily polluting vehicles. The aim of the LEZ is to improve the quality of the air in London.

14 *Using case studies*

Question

London is the largest city in the UK and can clearly make an impact in reducing carbon emissions in the country. Find out what plans your local authority has made in this respect.

Guidance

As with London, you could try to find out what your local authority is doing in terms of:

- subsidised, or free, home insulation
- improving energy efficiency in existing homes
- enforcing new building standards for energy efficiency
- investing in small-scale renewable energy schemes
- encouraging 'waste-to-energy' schemes
- providing efficient public transport
- promoting the purchase of low-carbon goods and services by council offices

For example, in Doncaster:

- The Save'n'Warm Inspiring Students programme aims to encourage young people (and their parents) to be more energy efficient and to be aware of energy issues. The Save'n'Warm Active Students scheme encourages students to take on more responsibility for energy efficiency in their schools. As a first step, energy monitors are nominated for the classrooms.
- On a larger scale, there is the A638 Quality Bus Corridor (QBC), which aims to encourage more people to travel by bus, with two new park-and-ride sites, dedicated bus lanes, and state-of-the-art buses and bus shelters along the routes.

Part 7

Other mitigation and adaptation systems around the world

Table 6.1 on page 52 illustrates that there is a wide range of mitigation and adaptation schemes and ideas around the world to address climate change, and many of these are on the sliding scale between the two categories of mitigation and adaptation. As with the previous section, these strategies operate in a range of contexts, from the global scale, to the national, regional and local scale. Some are already in place, while others are more speculative or even futuristic. This section examines a range of such strategies.

LULUCF

Land use, land-use change and forestry (LULUCF) is defined by the UN Climate Change Secretariat as 'a greenhouse gas inventory sector that covers emissions and removals of greenhouse gases resulting from direct human-induced land use, land-use change and forestry activities'. LULUCF has impacts on the global carbon cycle. These activities can add or remove carbon dioxide from the atmosphere, thereby contributing to climate change.

Land-use change is a large source of carbon dioxide and is thus an important contributor to climate change. The IPCC estimates that land-use change (e.g. conversion of forest into agricultural land) contributes a net 1.6 ± 0.8 Gt carbon per year to the atmosphere. In comparison, the major source of carbon dioxide, namely emissions from fossil-fuel combustion and cement production amount to 6.3 ± 0.6 Gt carbon per year.

There are six types of land-use change that can influence stocks of natural carbon storage:

- conversion of natural ecosystems (e.g. rainforests) to permanent croplands
- conversion of natural ecosystems (e.g. rainforests) for shifting cultivation

- conversion of natural ecosystems (e.g. rainforests) to pasture
- abandonment of croplands and pastures
- harvesting of timber (deforestation — Figure 7.1)
- establishment of tree plantations (afforestation)

When forests are cleared for conversion to agriculture or pasture, a large proportion of the above-ground biomass may be burned, rapidly releasing most of its carbon into the atmosphere. Forest clearing also accelerates the decay of dead wood and litter, as well as below-ground organic carbon. Local climate and soil conditions determine the rates of decay; in moist tropical regions most of the remaining biomass decomposes in less than 10 years. Some carbon or charcoal accretes into the soil carbon pool. When wetlands are drained for conversion to agriculture or pasture, soils become exposed to oxygen. Carbon stocks, which are resistant to decay under the anaerobic conditions prevalent in wetland soils, can then be lost through aerobic respiration.

Forest clearing for shifting cultivation releases less carbon than permanent forest clearing because the fallow period allows some forest regrowth. In general, the carbon stocks depend on forest type and the length of fallow, which vary across regions. Under some conditions, shifting cultivation can increase carbon stocks in forests and soils from one cut-and-regrowth cycle to another. Shifting cultivation usually has lower average agricultural productivity than permanent cultivation, so more land is required to provide the same products. In addition, shorter rotation periods deplete soil carbon more rapidly.

***Figure 7.1** Deforestation: one type of land-use change that can affect natural carbon storage*

TopFoto

Abandonment of cultivated land and pastures may result in recovery of forest at a rate determined by local conditions and, therefore, absorption of carbon back into the ecosystem.

Selective logging or harvesting (deforestation) may release carbon to the atmosphere through the indirect effect of damaging or destroying up to a third of the original forest biomass, which then decays as litter and waste in the forest. The rate of decay of the harvested wood is dependent on the end use — for example, fuelwood decays in 1 year, paper in no more than a few years and construction material in decades. The logged forest may then act as a sink for carbon as it grows back, and this will compensate gradually for the decay of the waste created during harvest. Clear-cutting of forest can also lead to the release of soil carbon, depending on what happens after harvesting. For example, harvesting followed by cultivation or intensive site preparation for planting trees may result in large decreases in soil carbon — up to 30–50% in the tropics over a period of up to several decades. However, harvesting followed by reforestation, has, in most cases, a limited effect ($\pm$10%). This effect is particularly prevalent in the tropics, where recovery to original soil carbon content after reforestation is quite rapid.

When tree plantations are raised on land that has been cleared specifically for afforestation, initially there are net carbon emissions from the natural biomass and the soil. The plantations then begin to fix carbon at rates dependent on site conditions and species grown. The timescale of carbon uptake in forest plantations depends on the nature and location of the plantations. Estimates include: 10 t ha^{-1} yr^{-1} for coniferous plantations in Australia and New Zealand, 1.5–4.5 t ha^{-1} yr^{-1} in coniferous temperate plantations of Europe and the USA, 0.9–1.2 t ha–1 yr^{-1} in Canada and Russia and 6.4–10.0 t ha^{-1} yr^{-1} in tropical Asia, Africa and Latin America.

Cropland soils can lose carbon as a consequence of soil disturbance (e.g. tillage). Tillage increases both aeration and soil temperature, making soil aggregates more susceptible to breakdown and physically protected organic material more available for decomposition. In addition, erosion can affect soil carbon stocks significantly through the removal or deposition of soil particles and associated organic matter. Soil carbon content can be protected and even increased through alteration of tillage practices (such as no-tillage or low-tillage regimes), crop rotations, residue management, reduction of soil erosion, improvement of irrigation and nutrient management, and other changes in forestland and cropland management.

Livestock grazing on grasslands, converted cropland, savannas and permanent pastures is the largest type of land use by area. Grazing alters ground cover and can lead to soil compaction and erosion (Figure 7.2), as well as alteration of nutrient cycles and runoff. Soil carbon, in turn, is reduced by these changes. Avoiding overgrazing can reduce these effects.

Croplands and pastures are also the dominant anthropogenic source of methane and nitrous oxide. Rice cultivation and livestock are the two primary sources of methane. Emissions of nitrous oxide are estimated to have increased significantly as a result of changes in fertiliser use and animal waste. Hence, alteration of rice cultivation practices, livestock feed and fertiliser use are potential management practices that could reduce sources of methane and nitrous oxide.

Figure 7.2 *Overgrazing of grasslands can lead to soil compaction and erosion*

There are three broad categories of LULUCF projects suggested by the UN Climate Change Secretariat:

- Reduction of emissions through conservation of existing carbon stocks — for example, avoidance of deforestation or improved forest management, including alternative harvest practices such as reduced-impact logging or restricting the use of fire.
- Carbon sequestration by the increase of carbon stocks — for example, afforestation, reforestation, agroforestry, revegetation of degraded lands, reduced soil tillage and other agricultural practices to increase soil carbon, or extended lifetimes of wood products.
- Carbon substitution — for example, use of sustainably grown biofuels to replace fossil fuels, or biomass to replace energy-intensive materials such as bricks, cement, steel and plastic.

CLIMATE CHANGE IMPACTS IN MALAWI

Case study 18

Background

Malawi is bordered by Mozambique, Zambia and Tanzania in the southern part of Africa, and covers 11.8 million hectares, of which 9.4 million are land; the rest is composed of water bodies, dominated by Lake Malawi. Of the total land area, 31% is suitable for rain-fed agriculture, 32% is marginal and 37% is unsuitable for agriculture. Malawi's population is estimated at 13 million, with a growth rate of about 2%. About

85% of the population is based in rural areas, with women forming 51% of the population and life expectancy as low as 40 years.

In tropical regions, the impacts of climate change are likely to be considerable. In general, developing countries are considered to be more vulnerable than more developed countries to the effects of climate change. This is attributed largely to a low capacity to adapt in the developing world. Of the developing countries, those in Africa are seen as being most vulnerable to climate variability and change. Like many other developing countries, Malawi (Figure 7.3) has not been spared from the severe impacts of climate change. In the last two decades, it has experienced a number of adverse climatic hazards. The most serious have been dry spells, seasonal droughts, intense rainfall, riverine floods and flash floods. Some, particularly droughts and floods, have increased in frequency, intensity and magnitude, and have impacted adversely on food and water security, water quality, energy and sustainable livelihoods of the most rural communities.

Rural communities in many parts of the country are currently experiencing chronic food deficits on a year-round basis due to the effects of floods and droughts. This situation has been compounded by the high prevalence of HIV/AIDS, which has created a large number of dependant orphans and has also impacted adversely on rural household food production systems, as well as on the quality of life and sustainable livelihoods. Women are particularly affected by drought because they are involved in water collection — shortage of water means that they have to travel long distances to collect it. The most impacted groups are the most vulnerable, female-headed households, children and the old, as well as those infected and affected by HIV/AIDS.

***Figure 7.3** The location of Malawi*

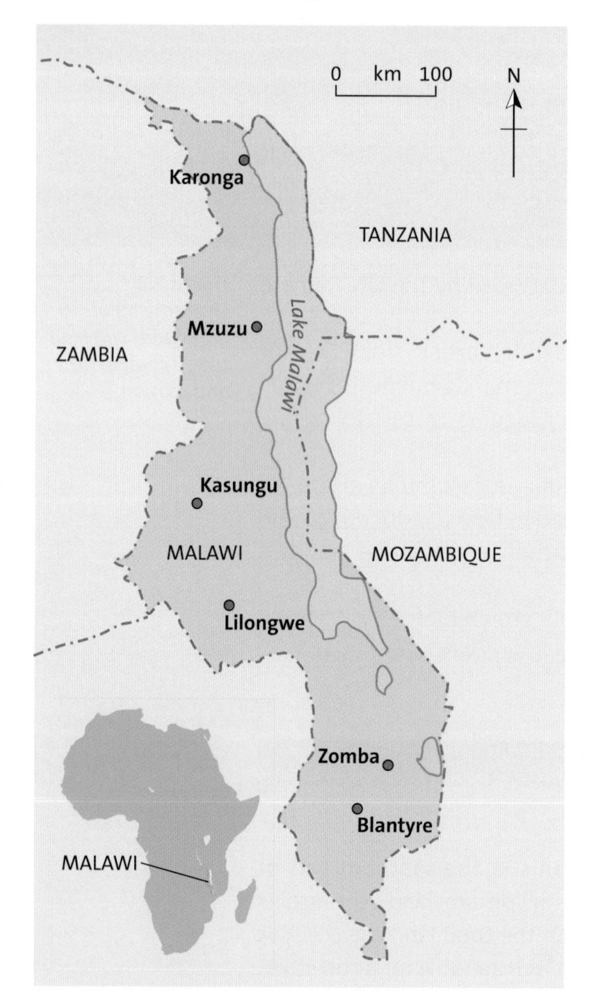

The increasing prevalence of recurrent floods and droughts is of major concern to the government of Malawi because of their far-reaching consequences on food, water, health and energy. Erratic rains have resulted in acute crop failure, despite huge efforts to improve seasonal weather forecasting at the beginning of the rainy season. Crop failure has resulted in food insecurity and malnutrition, particularly among vulnerable rural communities. Floods have resulted in the disruption of hydroelectric power generation, caused water pollution and increased incidence of diseases, such as malaria, cholera and diarrhoea.

Malawi's energy resources include biomass, coal, hydroelectric, solar and wind power. The energy system is dominated by biomass, which is the major energy source for domestic use and industrial development, with only 4% of the population having access to grid electricity. The energy sector is affected equally seriously by droughts and floods, which impact negatively on hydroelectric power generation along the Shire River, a major source of energy in Malawi. The water flow disruptions have been exacerbated by siltation caused by poor and unsustainable agriculture practices, deforestation, and noxious weeds such as water hyacinths.

Wildlife in Malawi has traditionally been treated by rural communities as a common good and, as part of the livelihood of a rural population of predominantly subsistence farmers and fishermen, has contributed to food security and household income. The major climatic hazard that affects the wildlife sector is drought. Drought affects animal reproduction and migratory habits. For example, the 1979–80 droughts resulted in the death of nyala antelopes in the Lengwe National Park in Chikwawa District and the migration of most animals from game reserves.

Malawi's water resources comprise both surface and ground waters. The surface water resources consist of a network of river systems such as Shire, Ruo, Bua, Rukuru and Songwe and lakes such as Malawi (the third largest in Africa), Chilwa, Chiuta and Malombe. The lakes and river systems cover up to 20% of the total surface area of the country. Water is a critical resource for human and industrial use, and for the maintenance of ecosystems. Increasing droughts and floods seriously disrupt water availability, quantity and quality. The city of Blantyre is quite often hit by water shortages, resulting in outbreaks of disease when drought is experienced. During floods, shortage of water is due to the bursting of water pipes and the silting of dams.

In terms of the forestry sector, the most important climatic threat is the extended droughts, which lead to land degradation and loss of soil fertility, as well as forest fires. For example, during the drought of 1995, some 5550 ha (or 36%) of Chongoni forest were destroyed by forest fires, resulting in smoke haze, pollution and loss of both seedlings and biodiversity.

National adaptation plans

The threat to food, health, water and energy posed by the extreme weather events of drought and floods has been the driving force for the preparation of Malawi's National Adaptation Programmes of Action (NAPA). The Malawi NAPA have identified the following priority areas of action to reduce the suffering of the most vulnerable communities:

- improving community resilience to climate change through the development of sustainable livelihoods
- restoring forests in the Shire valley catchment to reduce siltation and associated water-flow problems
- improving agricultural production under erratic rains and changing climatic conditions
- improving Malawi's preparedness to cope with drought and floods
- improving climatic monitoring to enhance Malawi's early warning capabilities and decision making and sustainable utilisation of Lake Malawi and resources in lake shore areas

To implement the above projects, Malawi requires US$22 million in aid to enable vulnerable rural communities and groups in targeted areas to adapt to adverse impacts of climate change.

Potential barriers to implementation

Malawi recognises the importance and urgency of addressing the problems associated with climate change because these affect the sustainable livelihoods of all Malawians — hence the need for the urgent implementation of the proposed adaptation options listed in the NAPA document. However, there are several barriers that could hamper the implementation of these activities. Apart from limited internal capacity to fund adaptation activities, Malawi is also constrained by several other factors, including:

- the extreme poverty of the most vulnerable groups, who are also illiterate, which makes it difficult to transfer new technologies and conduct meaningful long-term planning
- poor infrastructure, particularly poor roads and bridges, which makes it difficult to access rural areas, so there are difficulties in delivering farm inputs (e.g. fertiliser and seeds) and in accessing markets
- limited credit opportunities for rural communities to allow family households to access farm inputs easily
- food insecurity in the Southern Africa Development Community that would make it difficult for Malawi to acquire food from neighbouring countries, further aggravating the costs of coping with current droughts and floods
- the large number of HIV/AIDS orphans, which creates a major drain on family energy, cash and food; this is more critical among the poor in rural areas who have limited capacity to produce enough food and who are vulnerable to disease
- poor health conditions in rural communities, which lead to high rates of malnutrition, especially in children and the elderly, limiting the ability of the people to respond effectively to work opportunities

15 *Using case studies*

Question

Discuss the view that those countries contributing least to climate change are likely to suffer most from its impacts.

Guidance

The case study of Malawi can be used to provide specific detail in support to this question. A country such as Malawi has major problems adapting to climate change for the following reasons:

- It has a low capacity to adapt, lacking financial resources. International debt does not help in this respect.
- It suffers from two extremes of climatic impact — drought and flood, both of which lead to food and water insecurity. Drought can also lead to forest fires.
- There are problems with power generation and disease control.
- There is the additional problem of HIV/AIDS causing a huge burden.
- Within the country, it could be argued that women will be impacted upon most as they are the traditional collectors of water and firewood, as well as the carers.

Carbon offsetting

A carbon offset is a financial instrument that represents a reduction in greenhouse gas emissions. Although there are six primary categories of greenhouse gases, carbon offsets are measured in metric tons of carbon dioxide-equivalent (CO_2e). One carbon offset represents the reduction of one metric ton of carbon dioxide, or its equivalent in other greenhouse gases.

There are two markets for carbon offsets. In the larger market, companies and governments buy carbon offsets in order to comply with caps on the total amount

imagebroker/Alamy

Figure 7.4 Afforestation in Tibet

of carbon dioxide they are allowed to emit (see EUETS on p. 56). In 2006, about \$5.5 billion of carbon offsets were purchased, representing about 1.6 billion metric tons of carbon dioxide reductions.

In the much smaller voluntary market, individuals, companies or governments purchase carbon offsets to mitigate their own greenhouse gas emissions from transportation, electricity use and other sources. For example, an individual might purchase carbon offsets to compensate for the greenhouse gas emissions caused by personal air travel. In 2006, about \$91 million of carbon offsets were purchased in the voluntary market, representing about 24 million metric tons of carbon dioxide reductions.

Carbon offsetting as part of a 'carbon-neutral' lifestyle has gained some appeal and momentum, mainly among consumers in Western countries who have become aware of, and concerned about, the potentially negative environmental effects of energy-intensive lifestyles and economies. The Kyoto Protocol has sanctioned offsets as a way for governments and private companies to earn carbon credits that can be traded on a marketplace. The protocol established the clean development mechanism (CDM), which validates and measures projects to ensure that they produce authentic benefits and are genuinely 'additional' activities that would not otherwise have been undertaken. Organisations that have difficulty meeting their emissions quota are able to offset by buying CDM-approved certified emissions reductions. The CDM encourages projects that involve, for example, renewable energy production such as a wind farm, changes in land use and forestry, although not all trading countries allow their companies to buy all types of credit.

The commercial system has contributed to the increasing popularity of voluntary offsets among private individuals, companies and organisations as well as investment in clean technologies, clean energy and reforestation projects around the world (Figure 7.4). Offsets may be cheaper or more convenient alternatives to reducing one's own fossil-fuel consumption. However, some critics object to carbon offsets, and question the benefits of certain types of offset (see Case study 19).

As seen earlier, land use, land-use change and forestry (LULUCF) projects focus on natural carbon sinks, such as forests and soil. Deforestation can be avoided either by paying directly for forest preservation or by using offset funds to provide substitutes for forest-based products. There is a class of mechanisms referred to as REDD schemes (reducing emissions from deforestation and degradation), which may be included in a post-Kyoto agreement. REDD credits provide carbon offsets for the protection of forests and provide a possible mechanism to allow funding from developed nations to assist in the protection of native forests in developing nations.

Case study 19 FORESTRY IN THE UK AND CARBON OFFSETS

In April 2008, the UK Institute of Chartered Foresters held a debate to discuss the proposal that 'Forestry has a positive role in providing carbon offsets'. Table 7.1 summarises the views that were expressed both for and against the proposal.

***Table 7.1** Forestry's role in providing carbon offsets*

For	Against
At current levels of standing-timber prices and carbon-trading prices, the value of the carbon in a given volume of timber is approximately equal to the value of the timber itself	Some scientists believe that there will be a saturation of the carbon sink. It is a short-term solution compared with emission reduction
This means that for an owner planting a mainly spruce plantation on a 38-year rotation, or a native wood that will grow and stand for 100 years, sale of carbon credits could yield a potential additional income of approximately £1200 per hectare at the outset	Do forest carbon offsets encourage greenhouse-gas reducing behaviours in people or encourage business-as-usual emissions (or even increases)?
There are active buyers in the UK voluntary market at present, including businesses and organisations that are already reducing emissions by all practicable means and now wish to offset the remaining hard-core emissions in order to impress their customers, shareholders and staff	Forestry offsets are a distracting sideshow from genuine controls of greenhouse gas emissions, which are diverting attention from the real problem and allowing people to continue business-as-usual with climate-damaging consequences Forestry offsets are no more than 'greenwash' for high-emitting companies and individuals
A range of benefits will flow from the significantly increased level of afforestation that would result, including: ■ UK public and UK wildlife — a Newcastle University report in 2002 for the Forestry Commission valued the non-timber benefits of UK forests at over £1bn per year, of which the carbon element was approximately 10%, i.e. for every £1 invested in a carbon credit, £9 in additional public benefit could accrue over time ■ more supplies of timber for sustainable construction and wood for fuel — both are facing a shortfall in supply	Lots of reasons for afforestation, improved forest management and avoided deforestation (greater biodiversity, sustainable development etc.) but these should be used in addition to, not instead of, emissions reduction
It is essential that the forest sector can 'capture this carbon cash'	Yes — we can make money out of it — but does that make it the right thing to do? This does not make it a climate-friendly practice

WATER MANAGEMENT IN MELBOURNE (AUSTRALIA)

Melbourne is located in Australia within the state of Victoria (Figure 7.5). It is Victoria's capital city and the business, administrative, cultural and recreational hub of the state. Greater Melbourne has a population of 3.6 million. The city of Melbourne (CoM) is well known for a full calendar of events and is a major shopping and recreational destination. Major event locations include the Melbourne Cricket Ground, Melbourne Tennis Centre Precinct, Telstra Dome and the Crown Casino complex. CoM attracts over 1 million international visitors every year. It supports Melbourne's position as Australia's pre-eminent centre for arts and culture, education, fine food and dining, and shopping. CoM contains many natural features and recreational areas, including two major waterways (the Yarra River and Maribyrnong River), and a network of parks, gardens and sports fields, including the Botanical Gardens.

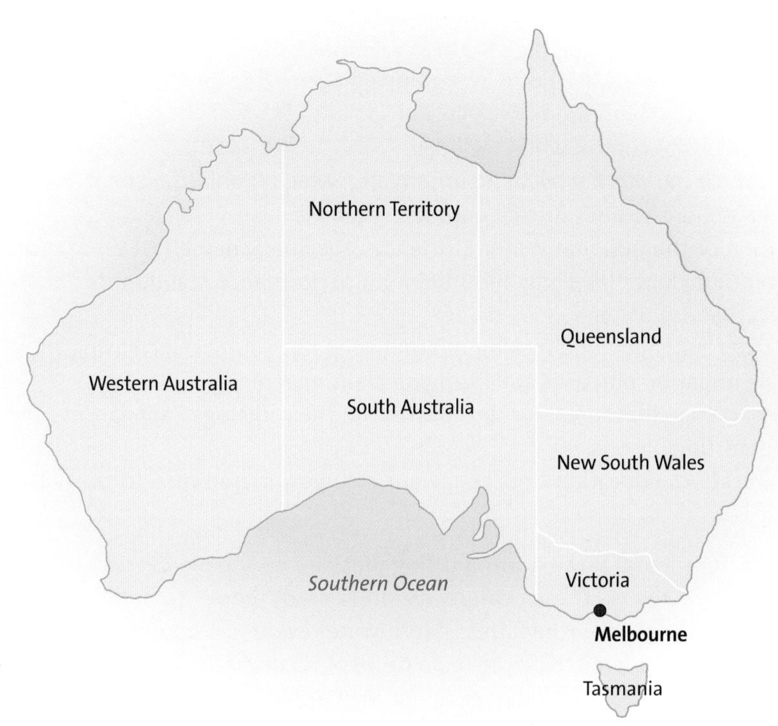

Figure 7.5 *The location of Melbourne*

Drought and reduced rainfall

Victoria is expected to become drier, with annual average rainfall decreasing by between 6% and 11% by 2070; most of that decrease is expected in the spring months. The projected decrease in rainfall and increase in evaporation will affect water supply. By 2030, the decline in annual rainfall, together with increased evaporation, is expected to cause less runoff into rivers, i.e. a potential decline of up to 45% in 29 Victorian catchments. For Melbourne, average stream flow is likely to drop by 3–11% by 2020 and by 7–35% by 2050. The latest regional climate change projections indicate that by 2070, runoff into the Yarra, Maribyrnong, Werribee and Bunyip rivers will decrease by up to 50%. Even with the current alternative sources of water supply, these reductions in water storage pose a threat to the water security of the affected area. Decreased water supply, along with warmer temperatures, is likely to increase drought risk and severity. As droughts become more severe, fire risk is expected to also increase. By 2020, the number of days with very high or extreme fire danger in Melbourne could be 10 or 11, compared with the current 9 days. The warmer, drier and longer summers expected in Victoria are likely to increase the frequency and intensity of bushfires.

Drought/reduced rainfall is more of an ongoing condition that Melbourne may attain with climate change, rather than a one-off event such as a heatwave or flood. A number of risks are considered critical and require immediate and ongoing management.

The most significant risk in the face of drought is insufficient water supply for Melbourne. The city of Melbourne has responded to the water scarcity risk by setting targets that seek to reduce water use by residents, employees and the council by 40%. To date, great savings have been made — a 48% reduction per employee, a 39% reduction per resident and 29% reduction in water use by the council. While it is positive that Melbourne residents are aware of the scarcity of potable water and have made significant reductions in consumption, with an increasing population further savings may be challenging. There is already a desalination plant planned for Victoria to produce 150 gigalitres (GL) of water annually from 2011, but further pressures on supply will need to be monitored closely and addressed.

Other methods to control water include:

- daily reservoir monitoring by Melbourne Water, which is published on its website
- water restrictions to limit further outdoor water use
- application of the national Water Efficiency Labelling Scheme (WELS) as a mandatory water-efficiency labelling and minimum performance standard for household products that use water
- parks water management seeking alternative water sources, increasing mulching, changing irrigation practices and changing planting regimes
- water-efficiency measures for businesses in their fittings, appliances and fire sprinkler testing
- communication, education and behaviour change programmes to prevent water wastage

A lack of environmental flows is also putting pressure on the biodiversity of the Yarra and Maribyrnong rivers. These waterways are already home to some endangered species and are at risk of further stress. Stormwater events, although providing water inflows, are also considered to be the main cause of toxins in the rivers.

The below average rainfall experienced in Melbourne over the past 11 years, with only 40% of average rainfall in 2007, has put significant strain on Melbourne's rivers. Low-flow conditions as a result of reduced water availability act as an environmental stress inducer, causing biodiversity impacts. Several drains in the lower reaches of the Yarra have high bacterial readings during dry weather. This has adverse effects on water quality. Deaths of large numbers of fish also occur during periods of low flow. The lower reaches of the Yarra are already subject to poor water quality and this is likely to be exacerbated with projections of further reduced rainfall over time. The Yarra is home to a number of native animals, such as frogs, platypus and fish, including many endangered species such as the growling grass frog and fish such as grayling, Macquarie perch and Murray cod.

Projects to protect the health of the Yarra by the Victoria State government include:

- works in the Yarra catchment to improve waterway and stormwater quality
- maintenance and upgrades of the sewerage system

- development and implementation of the Yarra Action Plan, initiatives of which include $930 000 being set aside to improve monitoring and communicating data on the health of the river. It involves a 3-year investigation programme with testing at 52 locations along the Yarra and its tributaries to track down likely sources of faecal pollution.
- weekly water quality sampling of *E. coli* at 12 sites on the Yarra River, which is undertaken by Melbourne Water's Yarra Watch Program
- monthly monitoring by Melbourne Water of water quality at 33 sites along the Yarra and its tributaries

The water restriction regime of Melbourne has helped to manage the significant drought issues of recent years. However, the heightened restrictions prohibit the irrigation of sporting grounds. In many areas of Australia, low rainfall levels and high evaporation have resulted in grounds hardening to the point at which there is an increased risk of serious injury. In response to this, CoM is planting drought-tolerant grasses, storing and recycling stormwater and investigating the use of artificial turf.

Overall, the future implications of reduced rainfall for Melbourne include:

- changes in the availability and quality of water supply available for residents and businesses
- increased cost of maintenance of public green spaces, parks and playing fields; it may also lead to the death of some vegetation and animals
- increased cracking of water supply, wastewater and stormwater pipes due to dry soils and tree roots seeking water; this leads to water wastage and stormwater leaking into sewers and causing overflows
- greater susceptibility of water stores to algal blooms (when combined with higher temperatures)
- disruption of agriculture, which would have an impact on food security for Melbourne

Extreme rainfall

Even though the overall rainfall for Victoria is expected to decrease, extreme rainfall events are projected to increase by 3% by 2050. For example, the intensity of the 1 in 20-year daily rainfall event may increase by between 5% and 70% by the year 2050. The future precipitation regime will have longer dry spells interrupted by heavier precipitation events, especially in the summer and autumn.

The future implications of extreme rainfall for Melbourne include:

- greater incidence of rainfall exceeding the coping capacity of existing stormwater and wastewater systems, leading to flash flooding. This would result in damage to infrastructure and property, as well as creating potential health risks from injury and contamination.
- higher concentrations of accumulated pollutants being flushed into metropolitan creeks, the Yarra River and Port Phillip Bay
- damage to infrastructure and buildings in areas vulnerable to landslide and erosion, especially in areas near waterways
- increased risk of disruption to key services due to flood damage — for example, electricity supply and transport
- higher road accident rates

The CoM's stormwater management system is characterised by old infrastructure, which until recently was poorly mapped and understood. Most drainage infrastructure in the CoM is over 60 years old, but some drains date back to the 1850s. In general, areas are protected for either a 10- or 20-year average recurrence interval. However, much of the existing drainage infrastructure is designed to accommodate 1 in 5-year events, and many road locations are not designed to adequately accommodate overland flow. There are several locations at risk of flash flooding in the CoM.

In response to these issues, since 2005, the CoM has implemented water-sensitive urban design as a stormwater management measure. The CoM states that this will continue to be a priority in order to meet stormwater quality targets to help improve the health of those waterways (Yarra, Maribyrnong and Moonee Ponds Creek) that have a history of poor water quality.

Question

Melbourne is taking the threat of climate change very seriously. Not only has it produced a detailed report on the prospect of future water supply, but it has examined a range of other issues in its *Climate Change Adaptation Report* 'Responding with resilience', which can be examined at: www.melbourne.vic.gov.au

Other issues it has examined are:

- **transport and mobility**
- **building and property**
- **social care, health and community**
- **business and industry**
- **energy and telecommunications**
- **emergency services**

Choose one of the above issues and investigate how this city (or any other you can find) is proposing to adapt to the challenges presented by climate change.

Case study 21 ZERO CARBON CITY RISES FROM THE DESERT

The United Arab Emirate of Abu Dhabi is building a $22 billion zero carbon showcase city for 50 000 residents. Masdar, to the southeast of Abu Dhabi, will be a three-tiered city with three decks that separate transport from residential and public spaces, and cars will not be allowed anywhere. On the lower deck, residents will be ferried around the city by thousands of personal rapid transport pods, which look like space-age buggies for four people. They are controlled by touch screens and guided by sensors in the ground.

About 6 m above the ground will be the main pedestrianised street level where businesses, shops and homes will be located. It will be vehicle free, except for more personal transporters and bicycles. Overhead, a light railway will run through the heart of the town and connect to Abu Dhabi.

Water will be drawn from dew and a solar-powered desalination plant. Most of the electricity for the 50 000 residents will be generated by solar panels on every roof and hung over narrow alleys where they will double as sun shades to keep the temperature low and reduce the need for air-conditioning.

Non-organic waste will be recycled. Organic waste will be converted into fuel for power plants. Dirty water will be processed and used to irrigate green spaces.

Overall, it is projected that Masdar city will need about a quarter of the energy of a normal city of its size. It will produce no waste, emit no carbon dioxide, and digital billboards will give a real-time account of energy usage and of how it is being generated. Their presence is a central part of the planners' efforts to instil in residents a sense of a living, breathing city. It will be completed by 2016.

MCDONALD'S BECOMES 'GREEN'

Case study 22

McDonald's opened a new 'green' restaurant in Chicago in 2008. While there are no plans to introduce a similar store in the UK, there are planned restaurants in Brazil, Canada and France.

The new site is designed to save energy. Some ideas are familiar, such as a system to capture rainwater. Others are more novel — for example, the table tops are made from recycled milk containers and one of the partitions is built out of Coca-Cola bottles. Some features are more obvious — for example, the use of skylights to let in more natural light and sending the cardboard waste for recycling. Other elements include reserved parking for hybrid vehicles, and LED signage to save on electricity costs.

It is estimated that the restaurant will use 25% less energy, thereby reducing utility bills. With 14000 restaurants in the USA and 31000 worldwide, the potential savings could be huge.

McDonald's UK arm has been experimenting with ideas to make its restaurants a little greener (Figure 7.6). In Sheffield, the fast-food chain has been converting waste into energy and 12 restaurants in Dorset took part in a trial to convert waste into compost for local farms.

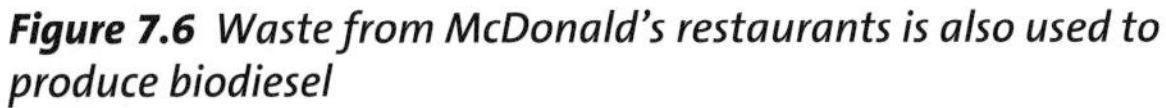

***Figure 7.6** Waste from McDonald's restaurants is also used to produce biodiesel*

17 Using case studies

Question

(a) Explain how pressure from consumers might persuade a business, such as a supermarket, to go 'green'.

(b) What steps could the supermarket take to achieve this?

Guidance

The market for green products and services is growing rapidly. In many countries consumer surveys report that growing numbers of people are willing to buy green products, if given the choice. For businesses, innovative product design and presentation combined with responsible marketing and communications could help ensure that this consumer interest translates into purchasing. However, the market for green products remains underdeveloped because people still find it difficult to locate products or trust their environmental claims. Businesses can help consumers to be more climate friendly, from the online click for carbon offsetting on a tourism booking website to the label on a product at the local supermarket.

Renewable energy sources

The use of renewable energy sources is one way in which mitigation can take place. All renewable energy projects are important, but the costs and benefits of each system must be weighed up. Large-scale renewable energy projects such as the Three Gorges Dam in China have a number of environmental negatives as well as the obvious benefit of providing large quantities of 'green' electricity. Biofuel crops provide renewable energy but take up huge areas of land that could be used for foodstuffs. In addition, deforestation in order to grow biofuel crops increases carbon dioxide emissions. Wind farms generate 'green electricity' yet cover large areas of countryside and coast. In terms of power generated per unit of land, photovoltaic cells, which generate electricity using sunlight, appear to be the most effective renewable energy source.

The following section examines a range of renewable energy sources at a range of scales and in a range of development contexts, and how they can combat, and cope with, the impacts of global warming.

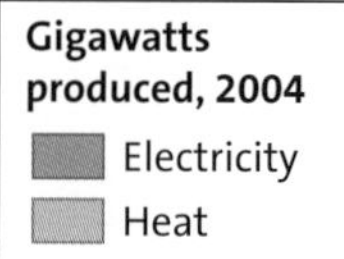

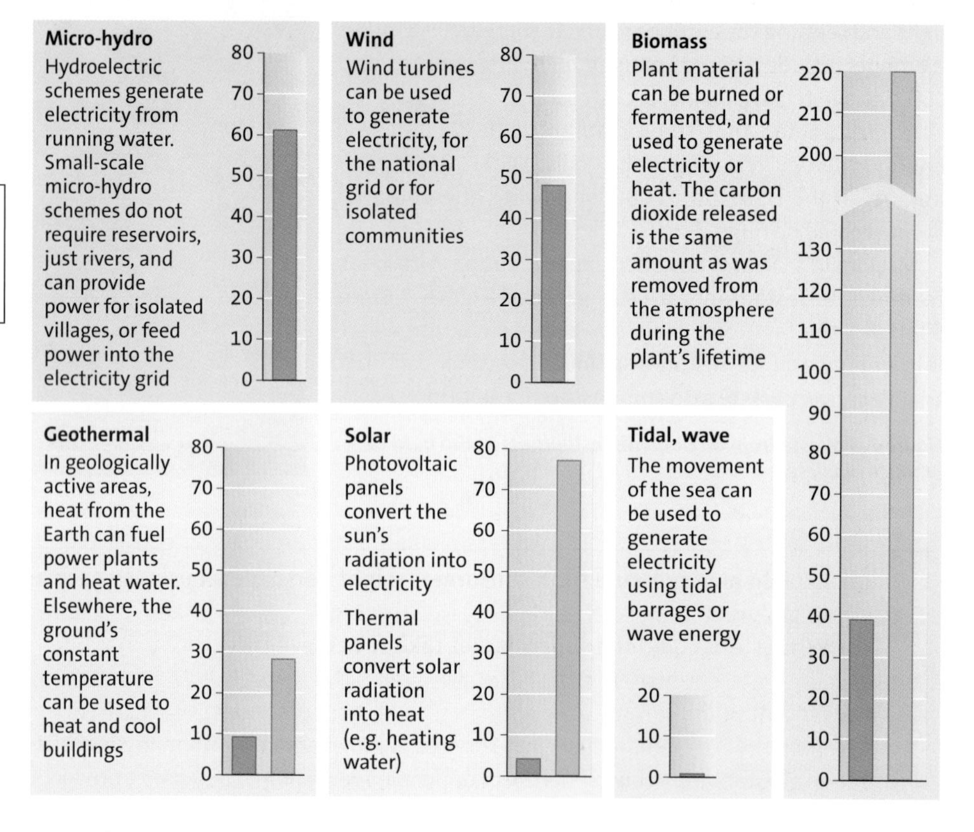

Figure 7.7 ***Renewable energy generation***

Source: *Atlas of Climate Change*

Solar power

In 2008, solar power contributed only 0.1% of the global electricity supply. This is largely because unit costs were high compared with conventional forms of electricity, but technical advances are now having a considerable impact on the cost of solar power. Solar electricity is produced currently in two ways: photovoltaic systems and concentrated solar power systems.

Photovoltaic (PV) systems

Photovoltaic systems convert sunlight directly into electricity using solar cells or panels, as in the modern-day calculator. A number of countries, including Germany (45% of the world's capacity), Japan, USA and Spain, have invested in large-scale schemes of this type. Several companies are now developing new forms of panels made from different materials, and which are much thinner. This allows them to be placed on and in materials such as glass and plastic. Entire roofs can be covered in this way. This thin-film technology is forecast to become much more important than traditional PV production.

Concentrated solar power (CSP) systems

These systems use mirrors, lenses and tracking systems to focus a large area of sunlight into a small beam. This concentrated light is then used as a heat source in a conventional power station. There are a number of CSP systems around the world, each differing in the way in which they trap and track the sun's energy.

Recent developments

- In March 2007, Europe's first commercial CSP plant was opened in Seville, Spain. It is hoped that by 2013 enough capacity will have been created to power 200 000 homes.
- In California, the Pacific Gas and Electricity Company has signed a contract to build three new CSP plants in the Mojave Desert by 2011. Steam will be used to drive turbines that will meet the electricity needs of 500 000 people.
- Australia is planning to build the world's largest single solar power (PV) plant in the state of Victoria by 2013.
- In the USA, there are ambitious plans to build a huge PV 'farm' covering 30 000 square miles in the state of Colorado. It is part of a plan to reduce US dependence on foreign oil and to cut greenhouse gas emissions by 2050. However, the scheme would create additional technological problems because new transmission systems would have to be created — existing transmissions systems would not be able to cope with the sheer power involved, and even if they could would lose too much power over the distances involved in carrying the electricity to market in the east of the country. Furthermore, as solar power systems do not operate at night, and demand for power is 24 hours, heat storage systems would have to be developed. To fund this 'solar grand plan' it is estimated that an investment of more than $400 billion will be needed over the next 40 years.
- In 2008, a similar scheme to the above was suggested at a conference in Barcelona. This would be based in the Sahara desert and would provide electricity to Europe. The cost and political implications of such a scheme are enormous. For example, to produce less than 10% of the world's power a solar 'farm' would require an area equivalent to a 100 km by 100 km square. Despite this, Algeria has begun the construction of a combined solar and natural gas plant, to begin production in 2010, and undersea cables to Sicily and Spain are being planned.

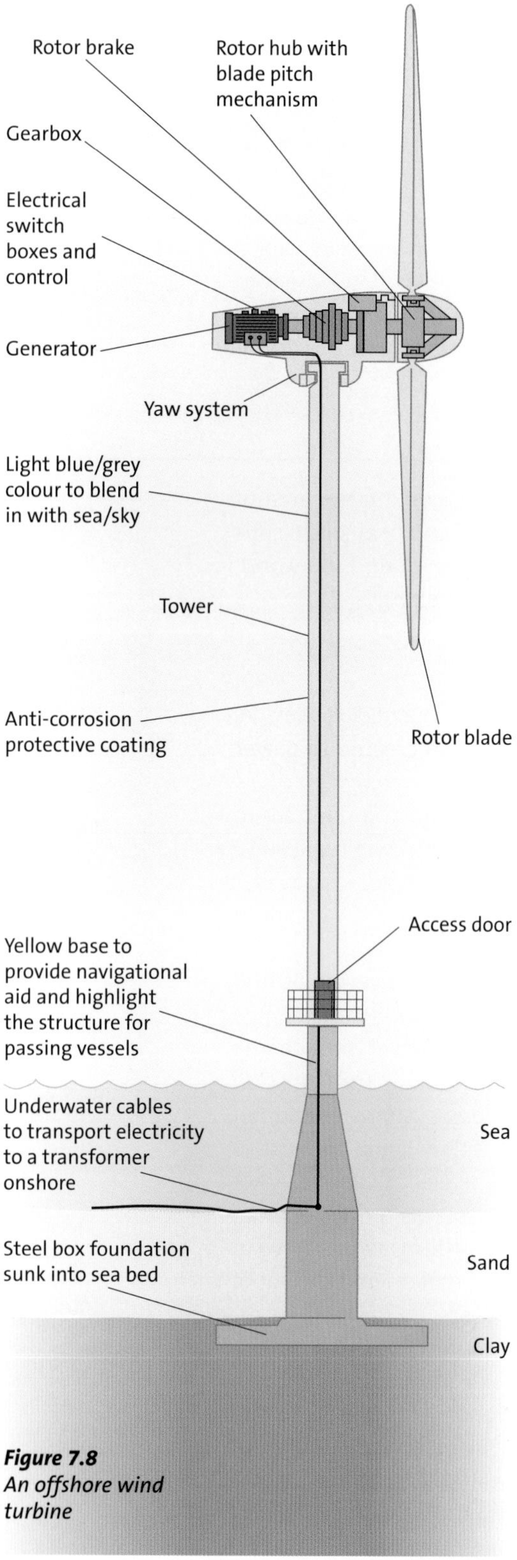

Figure 7.8
An offshore wind turbine

Wind power

Wind power is the fastest-growing renewable energy source. Many countries, particularly in Europe and North America, are seeking ways of developing wind power as one of their major sources of renewable energy.

Electricity is generated from the wind using a wind turbine. It is more economical if several turbines are sited in the same place in the form of a wind farm. As objections are sometimes raised to building wind farms in upland areas, wind energy companies have started to look for offshore sites, where larger capacity turbines can be used.

Wind energy is pollution-free and does not contribute to global warming. In Europe and North America, winds tend to blow strongly in winter when demand for electricity is at its highest. Although wind farms require large areas, the turbines themselves do not take up a lot of space (only 1% of the land on which they are sited), which allows farmland or natural habitats to exist around them. Electricity generation by wind energy is becoming increasingly competitive with coal-fired power plants and is cheaper than nuclear fuel. It is still not as cheap as gas-fired power stations, but wind energy costs are likely to go down in the future. Supporters of wind energy maintain that it represents an excellent example of sustainability.

Opponents of wind farms claim that many of the windiest sites are also areas of natural beauty. They argue that wind turbines are an unwelcome intrusion into the landscape and an eyesore. People's worries include:

- the noise the turbines create, particularly as wind farms are often sited in quiet locations
- the damage that they could inflict on wildlife (especially birds)
- the potential effect on property prices

Critics also point out that wind farms require large areas to produce only small amounts of energy. It can take over 7000 wind turbines to produce the same amount of energy as one nuclear power station. If wind energy is to be viable, a lot of wind turbines will have to be built.

WIND ENERGY PRODUCTION IN THE UK

Case study 23

The UK is the windiest country in Europe. In 2004, renewable energy sources in the UK generated just over 3% of the total electricity supply, 30% of which was derived from wind energy. The government's target is to generate 10% of the UK's total electricity supply from renewable sources by 2010 and it has now announced that it intends to increase this to 30% by 2020. Wind energy is probably best placed to meet these targets.

Although many wind farms are well established in the UK, there have been objections to both existing and planned developments. Even though he supports wind power, Jonathon Porritt, former director of Friends of the Earth, has his reservations:

> The real problem is that people building the things have been insensitive. They have put some of them in the wrong places and have not consulted local people or involved them in the benefits. The result is a growing anti-wind power lobby.

As a result of these objections, the industry is now locating some of its new developments offshore. The north Hoyle Offshore Wind Farm, off the coast of north Wales, has 30 turbines that generate enough electricity to power 50 000 homes. It has been estimated that this station will offset the release of 160 000 tonnes of carbon dioxide into the atmosphere every year. The naturalist David Bellamy is, however, campaigning against offshore wind farms, warning of 'plans that will make the British coastline ugly and impossible for birdlife'.

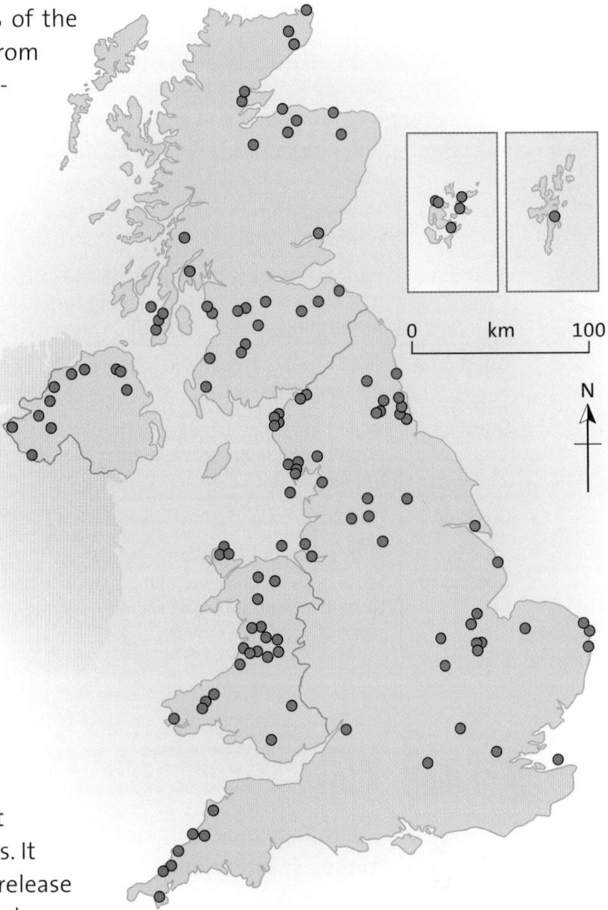

Figure 7.9 *Wind energy sites in the UK, 2007*

In the UK, there are over 200 operational wind farms producing 2735 MW onshore and 565 MW offshore. Further capacity for 17 000 MW is under construction, or going through planning applications. If all were installed, and assuming an average load factor (a measure of efficiency) of 0.3 (which some say is generous), they would produce the equivalent of 8% of UK-installed generating capacity.

There are, however, significant problems with offshore wind farms:

- The current design of wind turbines is such that a 5 MW turbine is 150 m tall. Several of these in an offshore farm create quite a sight, and are a navigational hazard.
- There is a severe shortage of vessels capable of erecting such turbines.
- Once erected, they are costly to maintain, and due to their location it is difficult to repair them should they break down (see p. 59).
- There are high costs of transmission back to land, which increase significantly with distance from the shore.

As described above, the installation of wind farms is not universally popular in the UK. You could attempt an exercise to assess the varying attitudes and perceptions of people to a wind farm being built in an area near where you live.

Guidance

Here is a range of questions you could use. All of these are closed questions. You might want to think of an open-ended question you could also ask.

Question 1	Have you yourself ever seen or visited a working wind farm within the UK?	Yes		No			
Question 2	Wind farms are necessary so that we can produce renewable energy to help us meet current and future energy needs in the UK.	Agree strongly		Agree slightly		Disagree	
Question 3	Wind farms are, or would be, ugly and a blot on the landscape.	Agree		Disagree			
Question 4	Wind farms are necessary so that we can produce renewable energy; what they look like is unimportant.	Agree		Disagree			
Question 5	I would be happy to have a wind farm in my local area.	Agree		Disagree			
Question 6	Would you consider installing a small wind turbine in your home in order to generate all or some of your own power?	Yes		No			

Case study 24 LOCAL-SCALE RENEWABLE ENERGY SCHEMES

There is a wide range of successful, local-scale, renewable energy schemes. Most of these use **intermediate technology** that is cheap and easy to build using local materials.

Improved stoves and household energy

In Sri Lanka, Kenya, Bangladesh and Sudan the introduction of improved stoves has been a great success. Since 1991, about half a million ceramic stoves have been produced and sold in Sri Lanka and many potters and installers have been trained in their construction and installation. It is estimated that future production will reach around 120 000 per year (Figure 7.10). In western Kenya women's groups make and sell around 11 000 ceramic stoves a year. As a result of their labours the women have gained status, self-confidence and financial independence. In Bangladesh, improved energy-efficient stoves using local materials consume less than half the fuel of a traditional open fire.

In Sudan, improved stoves using LPG were originally developed at Wau Nour, a camp for displaced and marginalised people on the outskirts of Kassala, as part of a project to reduce indoor air pollution due to cooking. These stoves are now being used in camps in Darfur, where their efficiency reduces the amount of firewood needed, which in turn reduces women's vulnerability when collecting fuel.

Micro-hydro power

This involves small-scale hydro schemes that generate up to 500 kilowatts (kW) of power. The micro-hydro station, which converts the energy of flowing water into

© Practical Action

Figure 7.10 *Improved stoves: since 1991 about half a million ceramic stoves have been produced in Sri Lanka*

electricity, provides poor communities in many rural areas with an affordable, easy-to-maintain and long-term solution to their energy needs.

These systems, which are designed to operate for a minimum of 20 years, are usually 'run-of-the-river' systems. This means that they do not require a dam or storage facility to be constructed but simply divert water from the stream or river, channel it in to a valley and 'drop' it into a turbine via a penstock (pipeline). This type of hydropower generation thus avoids the damaging environmental and social effects that larger hydroelectric schemes cause.

Cost for a typical micro-hydro system varies depending on the project. As a guide, every kilowatt of power generated costs around £800. A 6 kW system, enough to drive an electric mill and provide light for a community of about 500, would cost approximately £4800. As well as providing power for domestic lighting and cooking needs, village hydro schemes can also be used for charging batteries or for income-generating activities such as grain milling, depending on the needs of the community.

The Tungu-Kabri project, Kenya

The Tungu-Kabri micro-hydro Power Project is the first of its kind in Kenya. Funded by the United Nations Development Programme and developed by Practical Action East Africa and the Kenyan Ministry of Energy, the project benefits 200 households (around 1000 people) in the Mbuiru village river community, 200 km north of Nairobi (Figure 7.11). The project is cheap and sustainable, using small-scale technology that harnesses the energy of falling water to make electricity. It also alleviates the environmental problems associated with using wood and dung for cooking, diesel for milling and kerosene for lighting — and keeps on working, even in the face of drought.

Life is hard for the women and men in rural Kenya and the need for access to modern, 'clean' energy is acute. Ninety-six per cent of Kenyans live without access to grid

Figure 7.11 The Tungu-Kabri micro-hydro project, Kenya: debris has to be removed to prevent the turbine being damaged

electricity. In rural homes, families spend at least a third of their income on kerosene for lighting and diesel for the milling of grain. Kenyan women also devote a huge amount of time to collecting, processing and using wood and dung for cooking — time that could be spent on childcare, education or income generation.

The project now generates an estimated 18 kW of electrical energy. This amount will benefit about 200 households. In the months ahead, the villagers will be able to light their homes, save time and run small enterprises with this power. This will bring them a little money, to help buy clothes, food, and even schooling for their children. Water power also means that less wood is used, so the environment benefits.

Figure 7.12 A biogas plant on a small farm in Gurugodo, Sri Lanka

Biogas in Sri Lanka

With fuelwood becoming increasingly expensive and also scarce in some parts of Sri Lanka, there is a need to look for alternative cooking fuel. Cow manure and biogas technology provides a free, sustainable source of power all year round — and a better income for farmers. For on-farm biogas plants, cow dung is collected from specially adapted cattle sheds, mixed with water and channelled into fermentation pits (Figure 7.12). The resulting gas (65% methane) is produced as a by-product of this fermentation and is collected in a simple storage tank from where it is piped directly into the farmer's home to provide energy for cooking, laundry and lighting.

Incomes have risen as women and girls are freed from up to 2.5 hours a day of domestic labour (fuel collection, cleaning smoke-blackened utensils and disposing of animal waste), using the time instead for new income-generating activities. The biogas plants also produce a rich organic waste that is dried and used as fertiliser. Both fertiliser and fuelwood are increasingly expensive in the country and biogas has a potentially important future.

19 *Using case studies*

Question

Using examples, explain the phrase 'think global, act local'.

Guidance

Intermediate technology schemes as described above can be used to help explain this concept. There is a range of small-scale renewable energy projects including micro-hydro, wind turbines, biomass and solar power schemes that you could research and use in response to this question. Such schemes:

- develop eco-friendly and 'green' strategies that work with nature
- provide community-based solutions that facilitate locally developed initiatives
- provide pro-poor solutions, thereby helping the world's poorest people
- encourage more efficient use of resources so that they are conserved for future generations by not over-using them now

Community-based solutions work well because they are 'bottom-up' — developed by local people instead of being imposed by government. They are applicable to all aspects of mitigating and adapting to climate change in countries in all stages of development. For a community-based scheme in the UK, you could investigate **www.climatex.org**, which is based in Wolvercote in Oxfordshire.

Part 8

Climate change: the ongoing debate

The debate regarding climate change is likely to continue for years to come: whether it is happening, what will be the effects, how can it be reduced? Although there would appear to be a scientific consensus, there are still notable people who question its existence, the possible effects and the means to a solution. However, there would also now appear to be an increasing political consensus. Even the USA, which has dragged its feet in the early part of the twenty-first century, appears to be coming into line with the rest of the world. The United Nations Climate Change Conference in Copenhagen in December 2009 marks the latest stage of this political and scientific consensus.

This section seeks to examine the range of views on climate change that exists. The bulk of this book has examined issues from the assumption that the evidence from the IPCC is accurate, strong and valid — evidence that has been elucidated strongly and coherently by Al Gore in the film *An Inconvenient Truth*, which was sent to all schools in the UK in 2007.

The following case studies are a selection of views from other organisations and individuals.

Case study 25 GREENPEACE

The following is the stance taken by Greenpeace:

GREENPEACE

The world is warming up. Already 150 000 people are dying every year because of climate change and, within 50 years, one-third of all land-based species could face extinction. If we carry on the way we are now, by 2100 the planet will likely be hotter than it's been at any point in the past 2 million years.

But catastrophic climate change isn't inevitable. We know that climate change is caused by burning fossil fuels. The technologies that could dramatically reduce our dependence on fossil fuels — decentralised energy, renewables and efficiency, hybrid cars, efficient buildings — already exist and have been proven to work. If we start cutting our emissions now, using these ready-to-go technologies, then there is still a chance to avoid the most catastrophic impacts of climate change.

What we're lacking is real action. The (UK) government needs to put in place meaningful policies to urgently reduce emissions — and to act on them immediately. Under New Labour, carbon emissions have risen. The government is set to miss its own emissions targets. Whether through political cowardice or industry lobbying, the government is failing to put their words into action.

We're the last generation that can stop this global catastrophe, and we need your help.

DAVID BELLAMY

Case study 26

David Bellamy is Honorary Professor for Adult and Continuing Education, Durham University.

Figure 8.1 *David Bellamy*

Extracts from an 'opinion piece' written for the New Zealand Centre for Political Research, entitled *The Global Warming Myth*, June 2007

Am I worried about carbon-induced global warming? The answer is no and yes. No because there has been no sign of global warming in New Zealand since 1955, this year snow has fallen in Portugal for the first time in 52 years and three US states are united by the fact that they have recorded their lowest temperatures ever. Yes because it has become a political football that has lost its foundations in real science. What especially worries me is that if anyone dares to question the dogma of the global warming doomsters who repeatedly tell us that C not only stands for carbon but for climate catastrophe, we are immediately vilified as heretics or worse as deniers.

I am quite happy to be branded a heretic because throughout history heretics have stood up against dogma based on bigotry. I don't like being called a denier because deniers don't believe in facts. There are no facts linking the concentration of atmospheric carbon dioxide with imminent catastrophic global warming; there are only predictions based on complex computer models.

I offer two simple data sets that are already in the public domain. The most reliable global, regional and local temperature records from around the world display no distinguishable trend up or down over the past century. The last peak temperatures were around 1940 and 1998, with troughs of low temperature around 1910 and 1970. The second dip caused pop science and the media to cry wolf about a catastrophic ice age just around the corner. Our end was nigh! As soon as the temperatures took an upward turn in the 1980s the scaremongers changed their tune switching their dogma to imminent catastrophic scenarios of global warming all based on computer models some that were proved to be as bent as the hockey stick which no longer features in IPCC's armoury.

Back to the data, how can a 60-year cycle of changing temperature give any credibility to claims that carbon dioxide is causing an inexorable march towards a climate Armageddon? The concentration of carbon dioxide in the atmosphere has risen throughout this time frame, yet the temperature has gone up and down in a cyclical manner. How can this be explained unless there are other factors in control overriding the effect of this greenhouse gas? There are of course many to be found in peer-reviewed literature, solar cycles, cosmic ray cloud control and those little rascals El Niños and La Niñas all of which are played down or even ignored by the global warming brigade. As are the positive aspects of carbon dioxide in the growth of plants. Add to that the fact that since 1998 the world's average temperature has shown a tendency to fall not rise. This fact the 'warmers' play down by arguing that you need a 10-year period, or better still a 30-year period to register a convincing change.

Case study 27 GEORGE MONBIOT

George Monbiot is a journalist and environmental and political activist. He believes that drastic action, coupled with strong political will, is needed to combat global warming. Monbiot has stated that climate change is the 'moral question of the twenty-first century' and that there is little time for debate or objections to a raft of emergency actions he believes will stop climate change, including:

- setting targets on greenhouse emissions using the latest science
- issuing every citizen with a 'personal carbon ration'
- new building regulations with houses built to German Passivhaus (ultra high efficiency) standards
- banning incandescent light bulbs, patio heaters, garden floodlights, and other inefficient technologies and wasteful applications
- constructing large offshore wind farms
- replacing the national gas grid with a hydrogen pipe network
- a new national coach network to make journeys using public transport faster than using a car
- all petrol stations to supply leasable electric car batteries with stations equipped with a crane service to replace depleted batteries
- scrapping road-building and road-widening programmes, and redirecting their budgets to tackle climate change
- reducing UK airport capacity by 90%
- closing down all out-of-town superstores and replacing them with warehouses and a delivery system

Monbiot says that the campaign against climate change is 'unlike almost all the public protests' that came before it.

> It is a campaign not for abundance but for austerity. It is a campaign not for more freedom but for less. Strangest of all, it is a campaign not just against other people, but against ourselves.

Monbiot also thinks that the economic recession of 2008–09 can be a good thing for the planet:

> Is it not time to recognise that we have reached the promised land, and should seek to stay there? Why would we want to leave this place in order to explore the blackened waste of consumer frenzy followed by ecological collapse? Surely the rational policy for the governments of the rich world is now to keep growth rates as close to zero as possible?

While he does recognise that recession can cause hardship, he points out that economic growth can cause hardship as well.

See www.monbiot.com

PHILIP STOTT

Case study 28

Philip Stott is an Emeritus Professor at the University of London. He regards himself as a 'mitigated sceptic' on the subject of global warming. He has not published scholarly articles in the field of climate change, although he has published books on the subject. He has been critical of terms like 'climate sceptic' and 'climate-change denier'; he believes in a distinction between the science of climate change and what he asserts is the myth of global warming:

> The global warming myth harks back to a lost Golden Age of climate stability, or, to employ a more modern term, climate 'sustainability'. Sadly, the idea of a sustainable climate is an oxymoron. The fact that we have rediscovered climate change at the turn of the millennium tells us more about ourselves, and about our devices and desires, than about climate. Opponents of global warming are often snidely referred to as 'climate change deniers'; precisely the opposite is true. Those who question the myth of global warming are passionate believers in climate change — it is the global warmers who deny that climate change is the norm.

Stott is also critical of organisations like the IPCC:

> Climate change has to be broken down into questions: Is climate changing and in what direction? Are humans influencing climate change, and to what degree? Are humans able to manage climate change predictably by adjusting one or two factors out of the thousands involved? The most fundamental question is: Can humans manipulate climate predictably? Or, more scientifically: Will cutting carbon dioxide emissions at the margin produce a linear, predictable change in climate? The answer is 'No'. In so complex a coupled, non-linear, chaotic system as climate, not doing something at the margins is as unpredictable as doing something. This is the cautious science; the rest is dogma.

Stott's 'alternative charter for a sound energy policy' begins with: '(what) we need are strong economies that can adapt to climate change' and he proposes that the Kyoto Protocol be dropped because of 'its command-and-control economics which have no chance of working in the face of world economic growth, especially in the developing world'. He believes that the Kyoto Protocol is 'moribund politically'. Stott is concerned that the UK is failing to address its core energy needs, which must involve a mix of **clean coal**, gas and probably nuclear power. Stott also encourages development of energy infrastructure in the developing world. He sees the alleviation of energy poverty, along with the need for clean water, as two of the most urgent world issues. He regards most renewables as helpful (although he is critical of wind power), but only marginal to the core requirements of an advanced society.

Case study 29 BJØRN LOMBORG

Bjørn Lomborg is a professor at the Copenhagen Business School.

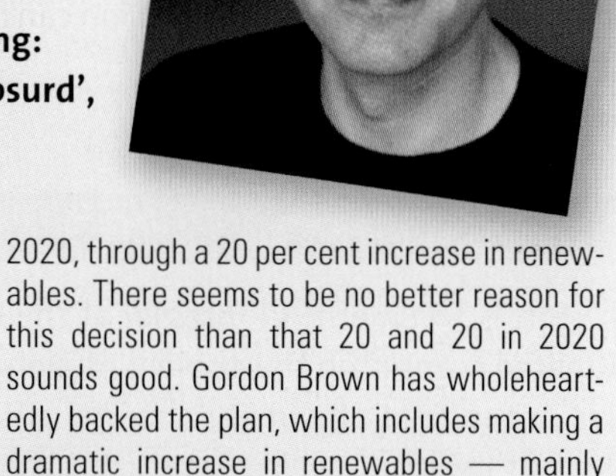

Figure 8.2 *Bjørn Lomborg*

Extracts from *The Times* 'Global warming: why cut one 3000th of a degree? It's absurd', 30 September 2008

Global warming is seen everywhere as one of the most important issues. From the EU to the G8, leaders trip over one another to affirm their commitment to cutting CO_2 to heal the world. What they do not often acknowledge — in part because it would lose them support — is that the solutions proffered are incredibly costly and will end up doing amazingly little good, even in a century's time. This is the truly inconvenient truth of the politics of global warming.

Let's be clear. I'm not contesting the existence of global warming. Doing so is silly, given the clear and strong results from the UN climate panel. Global warming will most probably warm the planet by between 1.6 and 3.8°C above current temperatures by the end of the century. However, we need to keep our cool: global warming's total cost will be only about one-half of 1 per cent of the net worth of the twenty-first century. Panicking is unlikely to lead to sensible policies. It could lead to exorbitantly expensive policies, which will do great harm.

It is a well-rehearsed point that the Kyoto Protocol was a terribly inefficient, hugely costly way to do virtually no good. Even if every industrialised country, including the United States, had accepted the protocol, and everyone had lived up to its requirements for the entire century, it would have had virtually no impact, even a hundred years from now. It would reduce the global temperature increase by an immeasurable 0.15°C by the year 2100. The cost of implementing Kyoto, taking the average figure from the various top macroeconomic models, would have been almost £100 billion annually for the rest of the century.

Yet the EU and others advocate that Kyoto-style policies are still right, only that much more than Kyoto is needed. The EU has promised to cut its emissions by 20 per cent by 2020, through a 20 per cent increase in renewables. There seems to be no better reason for this decision than that 20 and 20 in 2020 sounds good. Gordon Brown has wholeheartedly backed the plan, which includes making a dramatic increase in renewables — mainly 3500 wind turbines in the North Sea.

Computer modelling — using DICE (dynamic integrated model of climate and the economy) — shows that the net effect of the UK renewables effort is impossibly tiny. The temperature increase by 2100 without Mr Brown's plan would have been 2.4536181°C. With the best-case scenario the huge UK effort means that the temperature increase at the end of the century would be 2.4532342°C. The effect is a difference of about 0.00038°C — or about one three-thousandth of a degree in a hundred years. This is the equivalent of delaying the temperature increase by the end of the century by a little less than a week.

Of course, these numbers are way too precise: different models and assumptions would give somewhat different results. Yet because we are talking about relative change, the absolute climate sensitivity of the particular model matters very little. Thus the order of magnitude is robust and indicates an astonishingly small effect for a very large cost.

To make a simple comparison, the UN estimates that for about £40 billion annually, we could solve all major basic problems in the world — we could give clean drinking water, sanitation, basic education and healthcare to every person in the world. But instead we are spending a fortune achieving almost nothing.

When it comes to climate, we have to come to our senses. Yes, global warming is real and caused by human beings, but it doesn't mean we should panic in our policy decisions.

JONATHON PORRITT

Case study 30

The British population is 61 million, and will pass 70 million by 2028. The global population of 6.7 billion is expected to rise to 9.2 billion by 2050. Jonathon Porritt (a leading environmentalist) believes that the world should continue to focus on controlling population growth in order to reduce the effect of increasing population numbers on the environment, including the impact of global warming. He says that governments, including that of the UK, should improve family planning even if it means moving money from curing illnesses to increasing contraception and abortion.

The Optimum Population group, a campaign group of which Porritt is the patron, says that each child born in the UK will, during his or her lifetime, burn carbon roughly equivalent to 2.5 acres of oak woodland (an area the size of Trafalgar Square). He also chairs the UK government's Sustainable Development Commission, and states that:

> ...curbing population growth through contraception and abortion must be at the heart of policies to fight global warming. I am unapologetic about asking people to connect up to their own responsibility for their total environmental footprint and how they decide to procreate and how many children they think are appropriate. I think we will work our way towards a position that says that having more than two children is irresponsible.

John Robertson/Alamy

Figure 8.3
Jonathon Porritt

NICHOLAS STERN

Case study 31

Lord (Nicholas) Stern of Brentford who wrote the *Stern Review* in 2007 has stated that global warming should provide businesses with huge opportunities in the coming years and decades:

> Over the next 10–15 years the world is going to move strongly to low-carbon technologies. There is going to be a very rapid technological change. Areas like construction, transport and power are going to change particularly fast, and that's going to need huge investment as well as creating many businesses...The new technologies and investment opportunities of low carbon growth will be the main drivers of sustainable growth in the coming decades.

The scale of the challenge to business is huge. At the moment, humanity generates the equivalent of 50 billion tonnes of carbon dioxide a year, roughly equal to 8 tonnes for every person on the planet. There is, however, huge variation. Europeans generate 11–14 tonnes per head, Americans about 22 tonnes, while Africans generate typically 1–2 tonnes.

How big will this investment need to be? A report at the 2009 World Economic Forum held in Davos, Switzerland, has called for an annual global investment of £360 billion in green energy — including wind power, solar power, geothermal energy and biofuels. At the same meeting, the Chief Executive of Royal Dutch Shell (Jeroen van der Veer) stated that the key was to put a price on carbon dioxide emissions: 'there has to be a cost for emitting greenhouse gases,' he said.

Under such a global carbon market each country would be set a target for carbon dioxide emissions. If the target is exceeded, the country would have to buy allowances from a country that had a quota to spare. If it emitted less than its allowance, it could profit by selling the spare. Stern has estimated that under such a scheme developing countries could earn up to £70 billion a year from selling unused carbon dioxide allowances to Europe, America and other advanced countries.

Stern has acknowledged that the targets for cutting emissions are ambitious:

> It is going to be tough. We already have the equivalent of 420 parts per million of carbon dioxide in the air and we will reach 450 ppm in 8 years. Above that, and we only have a 50% chance of keeping global temperature rise below 2°C. The two great challenges of the twenty-first century, fighting world poverty and tackling climate change, must be tackled as an integral whole by a united world.

Case study 32 DAVID MACKAY

David MacKay is Professor of Natural Philosophy in the Department of Physics, University of Cambridge. In September 2009 he was appointed Chief Scientific Advisor to the UK Department of Energy and Climate Change.

Professor MacKay maintains that our present lifestyle cannot be sustained on Britain's own renewable energy resources unless we are prepared to cover huge areas of land and sea with wind turbines, tidal farms and solar cells. We can balance the energy budget either by reducing demand or by increasing supply or both. These reductions or increases must be big. Don't be distracted by the myth that 'every little helps'. If everyone does a little, then we will achieve only a little. He said: 'Obsessively switching off the phone charger is like trying to bail out the *Titanic* with a teaspoon.' We must do a lot.

Demand for power can be reduced by:

- reducing the population
- changing our lifestyle
- keeping our lifestyle but reducing its energy intensity through efficiency and technology

So, how about the latter? He says hydrogen-powered cars are a disaster — they are inefficient. On the other hand, electric-powered cars use ten times less energy and are efficient. Should we be investing in clean coal technology? Coal is a fossil fuel and Britain is estimated to have 7 billion tonnes of coal left. If we share this between 60 m people, we get just over 100 tonnes per head. Over a 1000-year period, this corresponds to 2.5 kWh of energy per day per person. In a power station performing carbon capture and storage, this sustainable approach to UK coal would yield only 0.7 kWh per day per person. Today, Britain's total energy consumption is 125 kWh per day per person. So, he concludes, clean coal is only a stopgap.

We could invest in nuclear fission. The amount of natural uranium used to provide the same amount of energy as fossil fuels is significantly small — as is the amount of waste. But, disposal of this waste presents a problem.

He states that we must revisit our attitude to big renewable facilities in the UK. For every nuclear power station we remove, we must find somewhere to put up 2000 wind turbines. He warns that for any renewable facility to make an appreciable contribution (i.e. against consumption) it has to be on a national scale. For example, if we were to grow crops for biomass — 75% of the UK would have to be planted; 4% of current energy from wave power would require 500 km of Atlantic coastline to be lined with wave farms; and wind farms the size of Wales would have to be constructed to make significant contributions — 50 kWh per day per person. But would the public accept and pay for this?

He states that we require a radical reduction in consumption, or significant additional sources of energy, or both.

RAJENDRA K. PACHAURI

Case study 33

Rajendra K. Pachauri is Chair of the IPCC and a joint winner with Al Gore of the Nobel Peace Prize, 2007. This is what he has to say (UNFCCC, Poznán, Poland, 1 December 2008):

> From the Fourth Assessment Report (of the IPCC) we now know the serious impacts of climate change, which would accrue as a result of inaction. We also know the nature of their worldwide implications.
>
> The differential nature of climate change impacts and the existence of other stresses leave the poor of the world particularly vulnerable. The ethical aspects of this reality need to be accepted in devising the implementing mitigation actions.
>
> Our collective record of mitigation of greenhouse gas (GHG) emissions has not been very inspiring. GHG emissions have grown, of course, since pre-industrial times, but there has been an increase of 70% between 1970 and 2004. Hence, the record of global action at mitigation has been very weak, even though the UN Framework Convention on Climate Change (UNFCCC) was agreed on in 1992. This record goes against the spirit and intent of the UNFCCC.
>
> Mitigation of emissions of GHGs has various merits and is in itself desirable and feasible in several respects. If global mean temperature increase is to be stabilised between 2.0–2.4°C, then CO_2 emissions must peak by 2015. The cost of such a stringent path of stabilisation of the earth's climate would be very modest, if at all a cost would be incurred. For instance, for this trajectory the cost to the global economy would at most be less than 3% of the global GDP in 2030. In fact there are so many co-benefits from such action that if these were to be fully accounted for then these might actually result in a negative cost, or a net increase in economic output and economic welfare.
>
> Large co-benefits of mitigation would include health benefits on account of lower air pollution at the local level, higher energy security, higher yields in agriculture, and greater employment opportunities. The record of those countries that have proactively pursued greater use of renewable energy, major improvements in energy efficiency have been able to increase employment in the economy.
>
> But even the trajectory of stabilisation described above would leave some serious problems in the nature of impacts of climate change. We would need to consider whether the effort to limit increase in global mean temperature to about 2°C would be adequate because sea level rise due to thermal expansion alone with this trajectory would be between 0.4–1.4 metres. Add to this the melting of ice bodies, and we would have serious effects of sea level rise on low lying coastal areas and small islands.
>
> My plea...would be to please listen to and reflect on the voice of science, and please act with determination and a sense of urgency. We in the IPCC do not prescribe any specific action, but action is a must.

20

Having read this selection of views, and having formulated your own views, try the following tasks:

(a) Draw a line with 'climate sceptic' on one end, and 'climate catastrophist' on the other end, and then place each of the preceding people/organisations along this line.
(b) Write a statement, of no more than 250 words, to try to convince a 'climate sceptic' that recent climate change is unprecedented.
(c) Write a statement, of no more than 250 words, to try to convince a 'climate catastrophist' that climate change is not something to be too concerned about.

Second time lucky?

The UN climate summit in Copenhagen (December 2009) represented a unique moment in modern history — when a worldwide agreement on global warming could have been reached.

The election of Barack Obama had brought a new sense of optimism in the world — for many he is seen as a 'green' president.

However, any agreement would need to balance how the burden of greenhouse gases will be shared. China is now the world's fastest-growing economy, and the greatest carbon dioxide polluter. Is it fair to ask China to cut emissions at the risk of economic slowdown when the USA industrialised with few environmental controls?

TopFoto

Figure 8.4 *President Obama*

President Obama had already shown his 'green' credentials, including:

- the appointment of a new special envoy for climate change
- car manufacturers being forced to produce more fuel-efficient vehicles
- individual states being allowed to set tougher standards for vehicle emissions
- federal government buildings to be made more fuel efficient
- a doubling of capacity for 'green' energy generation over the next 3 years
- an announcement in April 2009 to regulate carbon dioxide emissions

Around the world, politicians and environmentalists alike waited to see if the world's second largest carbon emitter and the planet's largest economy was finally able to take a lead on tackling climate change.

Then, for 2 weeks in the middle of December, some 20 000 United Nations bureaucrats, representatives of non-governmental organisations, world leaders and accompanying ministers and experts, descended upon Copenhagen for the Conference of the Parties to the UN Framework Convention on Climate Change (COP15). Their aim was to devise a substitute for the expiring Kyoto Protocol. In addition, small island nations and vulnerable coastal countries had been calling for a binding agreement that would limit emissions to a level that would prevent temperatures rising more than 1.5°C above pre-industrial levels.

The outcome has been generally viewed as being disappointing.

The Copenhagen deal

A motion was passed at the end of the conference to recognise the following deal, which was brokered by President Obama with the leaders of China, India, Brazil and South Africa:

- There would be no legally binding agreement to reduce carbon emissions.
- There is a need to limit global temperature rises to no more than 2°C above pre-industrial levels.
- That developed countries promise to deliver $30 billion of aid to developing nations over the next 3 years, and have 'set a goal of mobilising jointly $100 billion a year by 2020 to address the needs of developing countries' in coping with the impacts of climate change.
- The emerging nations (China, India, Brazil and South Africa) should monitor their own efforts to reduce carbon emissions and report to the UN every 2 years, with some additional international checks on progress.
- 'Various approaches' will be pursued regarding carbon markets and carbon trading.

The UN Secretary General Ban Ki-moon welcomed the climate deal in Copenhagen as 'an essential beginning'. He also said that the agreement must be made legally binding in 2010.

However, the deal failed to secure unanimous support, amid opposition from some developing nations. Several South American countries, such as Nicaragua, Venezuela, Cuba, Ecuador and Bolivia, were among a group saying that the agreement had not been reached through the proper process. African nations were not content either. Lumumba Di-Aping, the Sudanese negotiator, said the text 'asked Africa to sign a suicide pact'.

President Obama hailed the deal as unprecedented, but admitted that 'a deadlock in perspectives' had undermined the talks. 'We have much further to go,' he conceded. On the other hand, John Sauven, Greenpeace UK's executive director, said: 'There are no targets for carbon cuts and no agreement on a legally binding treaty. It seems there are too few politicians in this world capable of looking beyond the horizon of their own narrow self-interest.'

Research and fieldwork opportunities

Researching climate change locally is not straightforward. Change is difficult to measure — the impacts may be subtle, difficult to detect or notice, and require a long period of observation. Some would argue that such change, if it is indeed occurring, can only be measured using expensive and sophisticated equipment.

However, there are some sources of both primary and secondary data you could access.

Primary data

Phenology is the study of the timing of natural events, particularly in relation to climate. Examples include the dates of the first bluebell, or primrose, the first call of the cuckoo, or the date when hawthorn blossoms. These are then compared with records for other years. It is hard to predict climate change using such natural indicators, but some impacts are said to be occurring in some parts of Britain. It is also of interest that the impact varies between species — some, for example hawthorn, are showing signs of earlier growth, whereas others, for example ash and beech, are not.

It goes without saying that if you are using plants as indicators, you have to be able to identify them correctly.

Weather observation can be carried out over a short period of time (a week, or a month) and the readings compared with 'average' measurements using data from the Met Office (see below). A problem with this is that weather is highly variable.

Questionnaires are most useful for perception studies (as shown in Using case studies 18, page 84, where you ask people how they feel about an issue. Questions could include: 'How important is it that you take action to combat global warming?' or 'How important is it for the council to reduce its carbon footprint?'. The answers can be part of a bipolar exercise in which they are graded from 'very important'

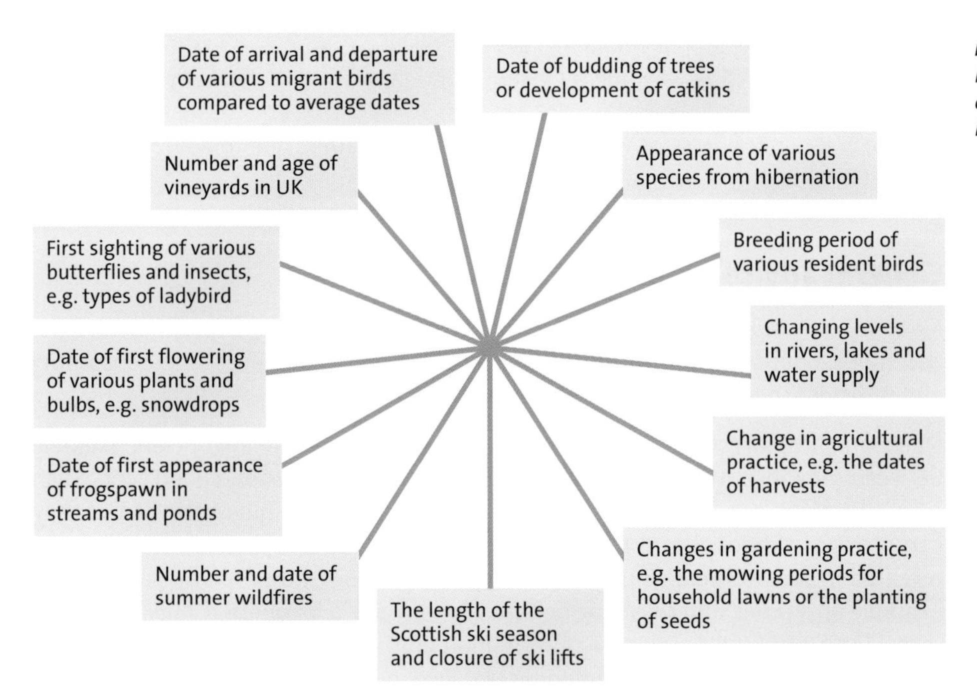

Figure 9.1 *Possible climate change response indicators*

to 'very unimportant', or in which respondents are asked if they 'strongly agree' or 'strongly disagree' with an idea or proposal, or somewhere in between (see general advice on questionnaires on page 100).

Other sources could include oral histories through interviews. You could ask local people what changes in the weather they have seen. Shop-owners may have noticed changes in the types of product sold, or the times at which they are sold. Sales of different foods and clothing may have started earlier or later depending on the item in question. Some supermarkets keep this information, which would act as a secondary source.

It may also be possible to record, photograph or video an extreme event such as a torrential downpour or flood.

Secondary data

There are a number of rich sources of secondary research data:

- **Newspapers** and other periodicals may carry a record of climate change, usually linked to extreme weather locally.
- **Internet blogs** are increasingly important for reporting on the attitudes towards local issues and events.
- The **Met Office** has a wealth of useful resources. Its website allows you to search through its database by location and climatic variable and access longer-term records, which may provide evidence for a changing climate.
- The **National Rivers Flow Archive (NRFA)** has records of the discharge levels of rivers. Flood events may be linked to climate change — shorter, more intense downpours appear to be becoming more common in the UK.

Other sources of secondary data include:

- photographs of past significant events or features that demonstrate climate change
- council records of loss of work days, or school closures, due to adverse weather conditions
- tourist board records of changes to visitor patterns

General advice on questionnaires

Many investigations at this level, and in this context of climate change, involve the use of questionnaires, the writing of which can be one of the most time-consuming and difficult aspects of an individual investigation.

Questionnaires are often seen as badly designed, with an insufficient balance between the more specific (closed) questions and the more open-ended type. Such open questions can certainly produce material that is less quantifiable, but they can provide additional information that is vital in trying to explain behaviour as revealed in the answers to the more direct questions. Questionnaire surveys, for example on studies into recycling habits, often involve a lot of effort but sometimes reveal little more than very basic information on the behaviour of people. Oddities in the pattern may not be accounted for because there was a failure to include open questions that may have gone some way to understanding people's motives for their behaviour.

Some questionnaires can become too sociological and therefore of limited geographical value, by not concentrating on the spatial aspects of the sample being studied. Questionnaires that represent little more than a social survey should be avoided, as they will give little scope for mapping and analysis of pattern.

When attempting a questionnaire survey there are a number of guidelines that can be followed:

- Keep it as simple as possible — busy people will not like answering too many questions.
- Try to write a mix of closed (yes/no answers or multiple-response) and open questions (choice of answers or free statements).
- Decide on an adequate sample size.
- Always try to introduce your questionnaire in the same way (write a brief introduction).
- Put the questions in a logical sequence.
- Ask questions that will produce data that can be analysed.
- Think carefully about sensitive questions and use tick boxes for such information as respondents' ages — it is often better to offer categories rather than insisting upon an exact figure.
- Try to ask questions about a person's behaviour, and not about how they perceive their behaviour.
- Pilot your questionnaire by testing it out to see if it will produce the material that you want.

- Always seek approval from your teacher or tutor before proceeding in order to avoid including insensitive questions, and also to prevent harassment of local people by swamping the area with too many questionnaires.
- Obtain a document from your school/college that states exactly what you are doing.
- Always be polite, look smart and smile, and do not get upset if people refuse to answer.
- If you intend to stand outside a specific service or in a shopping centre, it is advisable to seek permission.
- *Never* work alone.

21 Using case studies

Question

Your A-level group has been asked to conduct an enquiry to assess the impacts of climate change on your home area. Design an appropriate enquiry to meet the objectives of the task.

Guidance

You should follow this sequence:

- Identify key questions — Which area is to be studied? What time period should the enquiry cover? Which climate-related variables will be studied?
- Suggest a hypothesis/hypotheses — for example, flood frequency has increased/flowering is earlier/there have been changes in species diversity — each comparing the present with previous surveys.
- Identify the data required to test the hypothesis:
  - primary data — a fieldwork investigation using observations/measurement for which you will need to devise appropriate sampling techniques
  - secondary data — examine a range of different sources, for example data from agencies that have recorded changes (if any) over time (e.g. the Environment Agency for river discharge). Newspaper articles of notable weather events could be studied. You could make use of surveys such as BBC Springwatch.
- Present the evidence from the primary and secondary sources in map and/or graph form and use photographic evidence to illustrate and support the data.
- Data analysis — describe, analyse, evaluate and interpret the primary and secondary data. You could also suggest appropriate statistical means of analysing the data.

Index

A

B

C

H

I

J

K

L